Design and Construction of Pavements and Rail Tracks

BALKEMA – Proceedings and Monographs
in Engineering, Water and Earth Sciences

Design and Construction of Pavements and Rail Tracks

Geotechnical Aspects and Processed Materials

Editors

Antonio Gomes Correia

University of Minho, Guimarães, Portugal

Yoshitsugu Momoya

Railway Technical Research Institute, Tokyo, Japan

Fumio Tatsuoka

Tokyo University of Science, Japan

CRC Press is an imprint of the
Taylor & Francis Group, an **informa** business
A TAYLOR & FRANCIS BOOK

CRC Press
Taylor & Francis Group
6000 Broken Sound Parkway NW, Suite 300
Boca Raton, FL 33487-2742

First issued in paperback 2019

Typeset by Charon Tec Ltd (A Macmillan Company), Chennai, India

ISBN-13: 978-0-415-43362-4 (hbk)
ISBN-13: 978-0-367-38908-6 (pbk)

Visit the Taylor & Francis Web site at
http://www.taylorandfrancis.com

and the CRC Press Web site at
http://www.crcpress.com

Table of Contents

Preface

The International Technical Committee, TC 3 "Geotechnics of Pavements", of the International Society for Soil Mechanics and Geotechnical Engineering – ISSMGE, started in 2001 under the proposal of the ISSMGE Board 2001–2005. It follows the previous activities of ETC 11 (1997–2001) during which an important dissemination work was done culminated with three important publications: "Unbound Granular Materials. Laboratory Testing, In Situ Testing and Modelling" (Gomes Correia, 2000), "Compaction of soils and granular materials" (Gomes Correia & Quibel, 2000) and "Geotechnics for roads, rail tracks and earth structures" (Gomes Correia & Brandl, 2001).

This book intents to compile several selected contributions of the work carried out during 2002–2005 by TC3, complementary to the publication "Geotechnics in Pavement and Railway Design and Construction" (Gomes Correia & Loizos, 2004). It assembles recent knowledge in geotechnical aspects applied to pavements and railways dealing with:

- Geotechnical aspects related to foundation layers of pavements and rail tracks.
- Earth structures in pavement and railway construction – Promoting the use of processed materials and continuous compaction control.
- Strengthening and reinforcement of pavements and rail track.

These contributions represent an excellent source of state-of-art developments enhancing a wider application of geotechnics in design, construction, maintenance, upgrading of roads and railways. It also covers the related environmental aspects.

The editors would like to express their sincere thank to the TC3 members (2002–2005) for their contributions and to the President of ISSMGE, Prof. van Impe (2001–2005) for his continuous support and encouragement.

December 2006

A. Gomes Correia
University of Minho, Portugal; Chairman of TC3 (2002–2005), ISSMGE

Y. Momoya
Railway Technical Research Institute, JAPAN

F. Tatsuoka
Tokyo University of Science, JAPAN

REFERENCES

Gomes Correia, A.; Loizos, A. (Editors). Geotechnics in Pavement and Railway Design and Construction, MillPress, Rotterdam, Netherlands, Athens, December 2004.

Gomes Correia, & H. Brandl (Editors). Geotechnics for Roads, Rail Tracks and Earth Structures. Swets & Zeitlinger Publishers, Rotterdam, The Netherlands. 2001.

Gomes Correia, A & Quibel, Alain (Editors). "Compaction of Soils and Granular Materials". Presses de l' Ecole Nationale des Ponts et Chaussées, Paris, 2000.

Gomes Correia, A. (Editor). "Unbound Granular Materials. Laboratory Testing, In Situ Testing and Modelling". A. A. Balkema, Rotterdam, 1999.

1. Strengthening and Reinforcement of Pavements

1

Reinforcement of Pavements with Steel Meshes and Geosynthetics

H.G. Rathmayer
VTT – Building and Traffic, Geoengineering, Espoo, Finland

ABSTRACT: The European COST REIPAS research project (COST-348, 2002) has been initiated to enhance the use of reinforcement of pavements with steel meshes and geosynthetics. The basic common principles of reinforcing practice is laying the steel or geosynthetic mesh below the upper layers of the pavement either during the maintenance or reconstruction of existing roads or construction of new sections. The COST REIPAS project is taking a step towards practicable guidelines for the structural design of reinforced pavements and road sub-bases and to reach a consensus on the methods to determine relevant material parameters essential for analysing or predicting the behaviour of the reinforced structures. Design approaches developed either for the utilization of geosynthetic reinforcement materials or for the utilisation of steel grids are referred.

1 INTRODUCTION

In Europe road pavements have been reinforced with geosynthetics for more than 40 years, but also steel reinforcement is applied already for two decades to asphalt pavements and in the unbound granular base. Utilisation of reinforcing elements in road pavements is a multi-purpose solution to prohibit reflective cracking, to prolong the service life of pavements and thus to reduce maintenance costs and further more to increase resistance of the road structure against frost heave or differential settlements. The beneficial effects in terms of reduced construction costs and/or enhanced service lifetime have been verified both in research projects and in field trials. Reinforcing materials may be used both in the unbound granular base and in the asphalt overlay. The European COST REIPAS research project (COST-348, 2002) has been initiated to look into the use of reinforcement of pavements with steel meshes and geosynthetics. Based on the preliminary results from the COST REIPAS project an overview is given on typical applications and solutions, on the experience gained and also on ongoing research and development work. The terminology used in this paper is based on the recommendations from COST REIPAS (COST-348, 2005) and is shown in Figure 1.

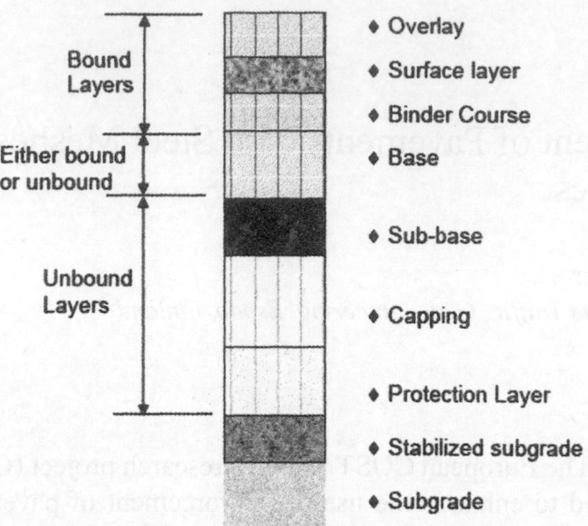

Figure 1. Terminology of pavement structure used in the COST REIPAS project.

Various solutions to place reinforcement between pavement layers or to modify the mechanical characteristics of pavement layers are presently available. The materials applied in the different technologies vary from steel grids, plastic grids and meshes to woven and non-woven geotextiles.

Reinforcement of pavements has several advantages, both economical, environmental and for traffic safety. The method is presumed to allow thinner road structures and longer life cycles, which lead to a saving in natural resources due to prolonged service intervals. Integration of reinforcing materials provides a cost-effective solution for rehabilitation and so a reduction in maintenance costs.

2 FUNCTIONS AND BENEFITS

The COST REIPAS project compiled the existing knowledge through a set of questionnaires amongst the experts in the participating countries. These resulted in descriptions of the damage cases for which reinforcement products are relevant and in an overview of assessment methods. In addition previous and ongoing research on the subject of reinforcement of pavements was reviewed and also the experience gained outside Europe, both in industrialised and developing countries. Further more the benefits that have been gained up to date by reinforcing of pavements and road bases are identified and the goals that should be set for the future development are formulated. The purpose of using reinforcement thus is to

- increase pavement fatigue life,
- minimize differential and total settlement,

- reduce rutting – surface and subgrade,
- prohibit or limit reflective cracking,
- increase resistance to cracking due to frost heave,
- reduce natural mineral usage,
- reduce maintenance costs,
- increase bearing capacity,
- enable bridging over voids,
- enable economic construction platforms.

The way reinforcement is used is to a large extent dependent on local conditions. Foundation, moisture regime, climatic and traffic conditions, types of granular materials, types of overlay, precipitation etc., all are influencing the structural solutions, the types of reinforcement to be used and what effects can be achieved.

3 APPLICATIONS

3.1 Unbound layers

Geosynthetics used in the unbound layers for reinforcement are polymer geogrids, geotextiles and geocomposites. In addition steel grids and meshes are used for some functions. The reinforcement is installed under and sometimes within the unbound base, subbase, capping and stabilized subgrade layers of a pavement. A summary of the functions, location of the reinforcement and type of reinforcement are presented in Table 1 (COST-348, 2005).

The reinforcement can be used both for construction of new roads and for rehabilitation and upgrading of existing roads. When used in new roads the reinforcement is commonly placed directly on the subgrade and the most common function is to effectively increase the bearing capacity of the soft subsoil by distributing stresses induced by the wheel loads over a wider area.

The beneficial effects are related to a reduced pressure being applied to the soft subsoil and hence less deformation during the construction period and less deformation (differential settlements and rutting) during the service lifetime. Generally the beneficial effects of the reinforcement are increasing with decreasing subgrade strength and increasing traffic loads. In cases with very soft subsoil a multilayer solution with reinforcement at the subgrade combined with a second or third layer up in the road structure is occasionally used. Also in areas prone to subsidence, e.g. old mining areas, reinforced structures are commonly used for bridging over voids.

Typically in areas susceptible to frost the old gravel roads might have frost susceptible material and very low bearing capacity in the thawing period. Before installation of the new pavement structure it is common to use a separating geotextile and a grid or mesh reinforcement on top of the old road before the new base layer and

Table 1. Function, location and type of reinforcement in unbound layers.

Function	Base course	Subbase course	Capping layer	Stabilised subgrade
Avoidance of Rutting	Polymer grids Steel meshes Composite polymer grids/geotextiles.	Polymer grids Composite polymer grids/geotextiles. Geotextiles	Polymer grids Composite polymer grids/geotextiles. Geotextiles	Polymer grids Composite polymer grids/geotextiles. Geotextiles
Increase of Bearing Capacity	Polymer grids Steel meshes Composite polymer grids/geotextiles. Geotextiles	Polymer grids Steel meshes Composite polymer grids/geotextiles. Geotextiles	Polymer grids Composite polymer grids/geotextiles. Geotextiles	Polymer grids Composite polymer grids/geotextiles. Geotextiles
Avoidance of Cracking due to Frost Heave	Steel meshes Polymer grids	Steel meshes Polymer grids		
Avoidance of Reflective Cracking in areas of road widening	Polymer grids Steel meshes Composite polymer grids/geotextiles. Geotextiles	Polymer grids Steel meshes Composite polymer grids/geotextiles. Geotextiles	Polymer grids Steel meshes Composite polymer grids/geotextiles. Geotextiles	
Avoidance of Fatigue Cracking	Polymer grids Steel meshes Composite polymer grids/geotextiles. Geotextiles	Polymer grids Steel meshes Composite polymer grids/geotextiles. Geotextiles	Polymer grids Composite polymer grids/geotextiles. Geotextiles	
Control of Subgrade Deformation		Polymer grids Composite polymer grids/geotextiles.	Polymer grids Composite polymer grids/geotextiles.	Polymer grids Composite polymer grids/geotextiles.
Bridging over Voids		Polymer grids Steel meshes Composite polymer grids/geotextiles. Geotextiles	Polymer grids Steel meshes Composite polymer grids/geotextiles. Geotextiles	Polymer grids Steel meshes Composite polymer grids/geotextiles. Geotextiles
Construction Platform	Not normally a base layer	Polymer grids Composite polymer grids/geotextiles. Geotextiles	Polymer grids Composite polymer grids/geotextiles. Geotextiles	Polymer grids Composite polymer grids/geotextiles. Geotextiles

an asphalt overlay is installed. The function of the reinforcement in such cases is to reduce the deformation of the old road structure and to distribute the remaining deformations over the whole width of the repaired road.

A wide range of materials are included in and under the unbound layers and they have many different functions and effects. Whilst all reinforcing materials provide some benefit to the pavement it is not possible to define these benefits from standard laboratory testing. A first step to a classification has been developed

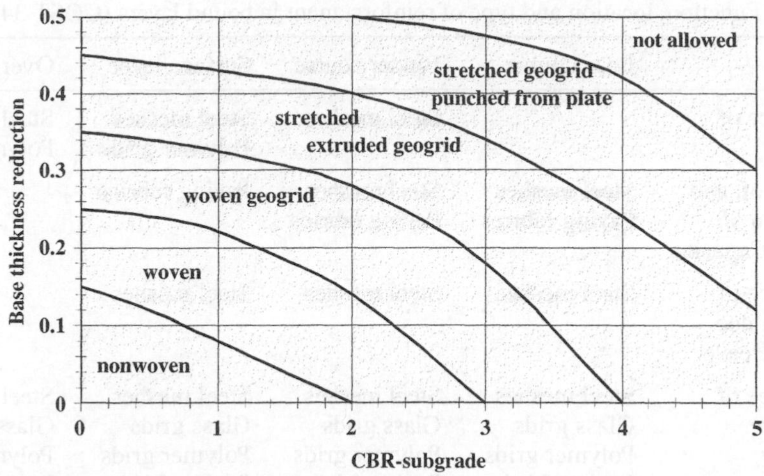

Figure 2. Example of classification chart for selection of geosynthetic reinforcement

in The Netherlands by CROW (CROW Publicatie 157, 2002) with the publication of the chart presented in Figure 2.

3.2 Bound layers

The reinforcement of the bound layers utilises a wider range of materials and addresses a number of problems. The types of reinforcement materials are geotextiles, polymeric grids, glass grids and geocomposites, but also steel meshes and steel grids. The use of reinforcement in bound layers is most commonly related to road rehabilitation and may both be for upgrading and installation of asphalt overlays on existing gravel surfaced roads or for repaving of existing paved roads with cracked overlay. A summary of functions, the location and type of reinforcement used is presented in Table 2.

For the purpose to avoid frost heave cracking special types of steel meshes are designed for applications which have road/lane width >5 m. In Table 3 recommended dimensions from Finland are given for different road classes. As the classification of roads varies between countries, these values are indicative.

4 FUNCTION MECHANISMS AND DESIGN

4.1 Unbound

The design of road structures includes a large variety of parameters; traffic loads, climatic conditions, drainage, subgrade and type of material in the pavement. Accordingly numerical models to describe the behaviour and theoretical design get complex and the inclusion of reinforcement does not simplify the task! Thus

Table 2. Function, location and type of reinforcement in bound layers (COST 348, 2005).

Function	Base course	Binder course	Surface layer	Overlay
Avoidance of Rutting		Steel meshes	Steel meshes Polymer grids	Steel meshes Polymer grids
Increase of, and protection of, Bearing Capacity	Steel meshes Paving fabrics	Steel meshes Paving fabrics	Paving fabrics	
Avoidance of Cracking due to Frost Heave	Steel meshes	Steel meshes	Steel meshes	
Avoidance of Reflective Cracking	Steel meshes Glass grids Polymer grids Paving fabrics	Steel meshes Glass grids Polymer grids Paving fabrics	Steel meshes Glass grids Polymer grids Paving fabrics	Steel meshes Glass grids Polymer grids Paving fabrics
Avoidance of Fatigue Cracking	Steel meshes Glass grids Polymer grids Paving fabrics	Steel meshes Glass grids Polymer grids Paving fabrics	Steel meshes Glass grids Polymer grids Paving fabrics	Steel meshes Glass grids Polymer grids Paving fabrics
Control of Differential Settlement	Steel meshes Polymer grids Paving fabrics Glass grids	Steel meshes Polymer grids Paving fabrics Glass grids	Steel meshes Polymer grids Paving fabrics Glass grids	

Table 3. Steel meshes as applied to different road classes.

Road/lane width (m)	Bar ϕ – spacing length dir. (mm)	Bar ϕ – spacing transv. dir. (mm)	Length of mesh (m)
12.5/7	8–100	6–150	13.00
10.5/7.5	8–150	6–200	11.00
10.0/7	8–150	6–200	10.50
8.0/7	7–100	5–150	8.50
7.0/6	7–150	5–200	7.50
7.0	7–150	5–200	7.50
6.0	6–100	5–150	6.50
5.5	6–100	5–200	6.00
4.0	6–150	5–200	4.50

design of reinforced granular bases is to a large extent based on experience and numerical models are trying to replicate what has been observed in the field.

For flexible pavements the linear elastic multi-layer mechanistic-empirical approach is widely used. In this approach the strain at the bottom of the overlay, the vertical stress on top of the granular base and the compressive strain at the top of the soil are the critical parameters.

The possible mechanisms how reinforcement functions in unbound granular base layers is outlined by A. de Bondt in reference (COST-348, 2005) for

- increasing the resistance against elastic deformations by increasing the horizontal stress level in the structure
- increasing the load bearing capacity of the pavement structure by distributing the load onto a larger area of the underlying soil
- reducing the mobilisation of the subsoil by reducing the shear stress transferred to the subgrade
- increasing the resistance to permanent deformation of the granular material itself by restraining horizontal movements of the granular particles (confinement).

The modelling should take into account both the effect of the elastic deformations of the pavement and the effect on the resistance against the plastic (permanent) deformations. Generally the effect of reinforcement in the granular bases is mostly related to the resistance against plastic deformations and to a less extent influences the elastic properties.

A pavement designer, who is interested in using reinforcement in a granular base layer, has to estimate the main effect of the function mechanisms. Thus modelling a reinforced structure has to take into account the effect

- on stiffness of the granular layer
- on the damage transfer function of the granular layer
- of reduced mobilisation of the subgrade.

Since the effect of a given reinforcement is highly dependent on the local conditions (traffic load, subgrade, materials and degree of flexure of the pavement) no general rules/guidelines for modelling of the effect are found. To a large extent design is based on producer specific empirically based design recommendations. However this commonly means that comparison between different solutions with reinforced base layers and comparison with more conventional solutions is very difficult. This also means that the use of reinforcement in the base layer has been taken only to a limited extent into account in general recommendations for road design.

4.2 Overlay

In case of asphalt overlay reinforcement is applied for new construction as well as for maintenance of old cracked overlays and in some cases to avoid reflective cracking over an old concrete overlay (de Bondt, 1999). Cracking can be caused by three different mechanisms:

- traffic loads,
- temperature variations over time,
- uneven soil movements (settlements, frost heave).

Two different function mechanisms are identified for the use of reinforcement in overlays:

- reduction of tensile strain in the asphalt by mobilization of tensile stress in the reinforcement,
- stress revealing interlayer to avoid transfer of tensile strain to underlying layers (geosynthetic materials).

Since the beneficial effect of the reinforcement highly dependent on the type of cracking mechanism, it is impossible to give general guidelines for design based on laboratory experiments. A design model or design guideline to be used for reinforcement in overlays has to take into account:

- dominating cracking mechanism,
- traffic characteristics,
- temperature variations,
- properties of the pavement,
- properties of the granular materials,
- conditions for existing pavement (in case of repaving),
- material properties for reinforcement,
- interaction reinforcement and surrounding overlay material (asphalt),
- construction equipment and procedures.

The questionnaire performed amongst the COST-REIPAS project members showed that only a small number of design models and procedures is available, but no one meets all the requirements which are mentioned above in a consistent way. Therefore, similar as for reinforcement in granular bases, design is to a large extent based on experience and is also product specific.

Also the design of maintenance treatments for cracked pavements, in which no reinforcement is included, has not got the necessary attention in the road construction community in the past. In almost all cases the selection of e.g. the mixture properties and the thickness of an asphalt overlay is based on empirical knowledge. This implies that relatively new options (such as e.g. geosynthetic and steel reinforcement) need a very long waiting period before they can be judged, which is unacceptable from an economical point of view. Several models and procedures are currently used in practise and are described more in detail in (COST-348, 2005).

4.3 Verification of effects

A number of field trials has been performed to evaluate the effect of reinforcement and also experience is available to a great extent. However only a limited number has been documented such that they can be used as reference.

As a part of the COST-REIPAS project also the methods used to verify the effects from reinforcement in pavements were investigated. The investigation included both verification of properties as basis for design and verification by testing in the field.

Unfortunately the FWD-test is not reflecting the beneficial effect of reinforcement and there is an obvious lack of suitable short term methods for that purpose.

5 EXAMPLES OF EMPIRICAL DESIGN PROCEDURES IN USE

5.1 Danish Road Design Procedure

The Danish road design manual offers two possibilities to integrate geosynthetic reinforcement:

a) a catalog for the selection of typical secondary pavements, chosen when the examination of the subsoil conditions is not sufficient,
b) a diagram for more subtle balancing of cost-effective material conditions, traffic loads and the actual subsoil.

According to the diagram method, the E-value (stiffness) of the subsoil has to be determined by means of triaxial tests in laboratory or by field experiments as e.g. the CBR-test (and making use of the well-known equation E in MPa is roughly 10 times the CBR in %). The extent and accuracy of the subsoil investigations must match the importance of the project, as the road will never be better than the quality of these investigations.

Geosynthetic reinforcement is implemented in the design procedure in the case the E-value of the subsoil is less than 30 MPa. The implication of using a geotextile allows the E-value to be multiplied by 1.8 and the increased value for the soil support can then be inserted into the design chart. The factor 1.8 originates from monitoring trial sections in the USA and is valid when typical non-woven fabrics are utilized (Steward et al., 1977).

5.2 REFLEX Design Model

REFLEX is the acronym for the BRITE/EURAM III research project "Reinforcement of Flexible Road Structures with Steel Fabrics to Prolong Service Life" carried out within the EC 4th Framework Program 1999–2002. The final report of the eight task groups is available via the web address http://www.vti.se/reflex/.

For the integration of a steel reinforcement in the design the multi-layer linear pavement structure modelling approach is used. The steel net reinforcement is included into the model as an additional layer within the pavement. As the reinforcement is described as one layer of the model, there are only three variables that can be changed so as to reproduce the effect of reinforcement on the resilient (elastic) deformation response of the pavement structure. These three variables include:

- the stiffness of the layer that represents the reinforcement
- the thickness of the layer that represents the reinforcement
- the interface properties of the layer as it is interacting with the layers above/below.

The basic idea of the "equivalent layer"-concept is to substitute the reinforcement with a layer of finite thickness of some tens of millimetres and to give that layer some sort of average properties of the steel and the surrounding material.

5.3 MSU/SINTEF Design Method

A comprehensive research project GeoRePave has been executed by MSU (Montana State University, USA) and SINTEF (Trondheim, Norway), see (Perkins, 2004). The objective of this project was to develop design methods for reinforced unbound base course layers in roads. It included the development of numerical material models and numerical modelling methods for road foundations. This research project has resulted in a proposed design procedure.

The motivation for including reinforcement in unbound base course materials is to reduce both construction and maintenance costs. The latter should be evaluated in terms of life cycle costs, which can be split into:

- construction costs including materials
- maintenance costs over the road lifetime
- environmental impact from the use of construction materials

The cost reduction due to the use of reinforcement may be expressed with so-called benefit ratio's and these may be evaluated in terms of reduced required thickness of the unbound layer or increased number of traffic passes before maximum allowable deformation of the road is exceeded. These are named the

- Base Course reduction Ratio (BCR) (=ratio for the reinforced base thickness) &
- Traffic Benefit Ratio (TBR) (=ratio of allowable traffic passes for a reinforced base course).

If the TBR ratio is applied, the reduction in maintenance costs in the future must be larger than the cost for purchasing and installing the reinforcement. The life cycle cost may however be the easiest to determine on basis of the BCR, since this compares designs for the same traffic load (same number of passes before the design criteria on allowable traffic is reached).

To be comparable with the unreinforced design, the design method for reinforced base courses must include parts of the experience and empirical relations derived from field tests. It is therefore a difficult task to compare the design of reinforced and unreinforced pavements in engineering terms. Design of reinforced pavements has thus to be performed in relation to conventional road design.

Starting from a 2D-axial symmetric Finite Element Model the unreinforced structure is dimensioned. The effect of compaction, traffic load and reinforcement on the horizontal and shear stresses is calculated stepwise. Due to its complexity, this procedure needs to be incorporated into a software/design package in order to be used on a routine basis in practice. The research in the GeoRePave project has provided the outline of the methods and is published on the web address http://www.coe.montana.edu/wti/wti/display.php?id=89.

6 CONCLUSIONS

Pavement reinforcement with geosynthetics has been used in Europe for more than 40 years, steel reinforcement more than 2 decades. It is obvious that use of reinforcement in road pavements has a promising potential and the beneficial effects of the reinforcement may both reduce construction costs and enhance the road performance. However, despite the large amount of research projects and of successful projects in the field with good experience, pavement reinforcement is still not recognised as a solution at the same level with conventional methods. This is to a large extent due to the lack of technically sound models for the function mechanisms of the reinforcement and proper non product related design models.

Currently general road design is to a large extent based on semi empirical methods and this complicates the inclusion of new materials and methods. A number of research projects has been carried out in order to develop models and methods and this experience is updated in the multinational COST-REIPAS project.

REFERENCES

COST-348 (2002): MOU for the implementation of a European Concerted Research Action Designated as COST Action 348. "Reinforcement of pavements with steel meshes and Geosynthetics" (REIPAS).

COST-348 (2005): Reinforcement of pavements with steel meshes and geosynthetics. (REIPAS) Draft Reports of WG 1–4. (to be published 3/2006).

CROW Publicatie 157 (2002): Dunne asfaltverhardingen: dimensionering en herontwerp.de Bondt, A.H. (1999) "Anti-Reflective Cracking Design of (Reinforced) Asphaltic Overlays". Ph.D.-Thesis, Delft Univ.of Technology (http://www.ooms.nl/adebondt/adbproef.html).

Perkins S.W, et al. (2004) "Development of Design Methods for Geosynthetic Reinforced Flexible Pavements". FHWA, Report Number DTFH61-01-X-00068.

Steward, Williamson, and Mohney (1977). "Guidelines for Use of Fabrics in Construction and Maintenance of Low-Volume Roads". USDA Forest Service, Portland, Oregon.

2. Geotechnical Aspects Related to Foundation Layers of Pavements and Rail Tracks

2

Improved Performance of Ballasted Tracks

A.M. Kaynia
Norwegian Geotechnical Institute, Oslo, Norway

D. Clouteau
Ecole Centrale de Paris, Châtenay Malabry, France

ABSTRACT: This paper presents an overview of the results of the research project Sustained Performance of Railway Tracks, SUPERTRACK. The project was funded by the European Community during the period 2002–2005. The main objective of the research was to enhance performance of ballasted tracks by better understanding the geomechanical behaviour of the elements of the track and their interaction. The paper presents the main findings of the research and assesses performance of a grouting method for track retrofitting.

1 INTRODUCTION

The majority of railway tracks in Europe rest on ballast. Low speed trains with speeds of around 200 km/h or less have been operating for a long time on such tracks without any major problem. High speed trains (HST) with speeds more than 300 km/h, have been operating on certain segments. Observations on railway performance have indicated that track problems, such as settlement and deterioration, tend to increase with train speed and lead to need for more maintenance. To make the HST more competitive, it is necessary to reduce the maintenance costs. The maintenance operations of a railway company represent about 6% of its annual turnover. This corresponds to a mean value of 15000 Euros per km per year. Among them 10% concerns the maintenance of the geometrical quality of the lines, which is mainly realised by tamping operations. Reducing the global volume of tamping operations could largely reduce the corresponding costs, and increase the availability of the lines.

To address the above issue, the research project SUPERTRACK, Sustained Performance of Railway Tracks, was funded by the European Community under the Competitive and Sustainable Growth Programme in the period 2002–2005. The main objective of the research was to enhance performance of ballasted tracks and reduction of maintenance by better understanding the geomechanical behaviour of the elements of the track and their interaction. To achieve these objectives the following research activities were performed:

i) Field testing and measurement
ii) large-scale material and model testings
iii) Development of numerical models for non-linear track response
iv) Implementation of track retrofitting for improved performance

The study was carried out jointly by a research group consisting of Norwegian Geotechnical Institute (NGI), Société Nationale des Chemins de Fer (SNCF), Administrador de Infraestructuras Ferroviarias (ADIF), Géodynamique et Structure (GDS), Centro de Estudios y Experimentacion de Obras Publicas (CEDEX), Ecole Centrale de Paris (ECP), Linköping University (LU), and Swedish National Rail Administration, Banverket (BV).

The following presents a summary of the research work and the main findings of the study. More details can be found in Kaynia (2006) and in the individual research reports of the project as well as the papers listed in the bibliography.

2 FIELD TESTING AND INSTRUMENTATION

Extensive field testing and instrumentation were made at four test sites in France and Spain during the research period. The test sites were selected with different objectives. The following gives a description of these sites, the measurements and the use of the collected data.

2.1 Tests at Zoufftgen, France

Zoufftgen is located in Lorraine, on the Metz-Luxembourg railway line, 4 km south from the border of Luxembourg. The measurement site was at the level crossing #10, on the road D56 between Kanfen and Zoufftgen, at kilometric point 200.361 on track #1. The geotechnical characteristics of the site were determined from a combination of lab tests and SASW measurements.

During the last years, tamping operations were not sufficient to guarantee a durable geometrical quality. The ballast layer was fouled by clay in many places, experiencing pumping and severe attrition. In October 2003, a complete renewal of ballast and under-ballast layers as well as replacing the sleepers and rails was undertaken in order to restore the track quality. The site was instrumented to measure the following response parameters: accelerations on sleeper, on ground surface and inside the track, deflection of the rail and rail pad, pressure under the ballast, and the wheel load. One of the objectives of the measurements at this site was to provide continuous data on the vibration and settlement of the track with time. In addition to providing substantial data for verification of the numerical models developed in this study, the measurements confirmed a sustained performance of the track with respect to the rail and rail pad deflections (1.7 mm and 0.5 mm, respectively) and distribution of the loads along the sleepers, ranging from 50 to 170 kPa.

2.2 Tests at Guadalajara site, Spain

A straight section on the Madrid-Zaragoza high-speed line was selected near Guadalajara, around K.P. 69+500. The test section comprised three zones two of which were on embankments, and the third was in a trench. Within the embankment, two line sections were chosen for instrumentation, Section 1, reaching about 18 m height, and Section 2, modelled in real scale in the track box (see Sec. 3), reaching about 4 m height. Section 3 was located in trench zone. Sections 2 and 3 were further divided into three subsections. Figure 1 shows a plan view of the selected test sections. The instrumentation was designed in order to measure the following response quantities: stresses at layer beds, contact stresses under sleeper, rail stresses, track deflections and settlements, track and ground vibrations, temperature and water content, and meteorological parameters. The track construction was completed in November 2003 and measurements continued throughout the research period.

The objective of these measurements was two fold: i) to provide data for short-term and long-term response of the track for different constructions, and ii) calibration of the measurements in the track box in order to predict the non-linear long-term track behaviour.

Figure 2 presents an example of data obtained by laser measurements. The figure shows the time history of the absolute displacements of the rail under a Talgo train one year after the track construction. According to this figure, and similar measurements, the rail displacements produced by the passage of trains at 200 km/h was in the order of 2.5 mm. The corresponding values after the track construction were

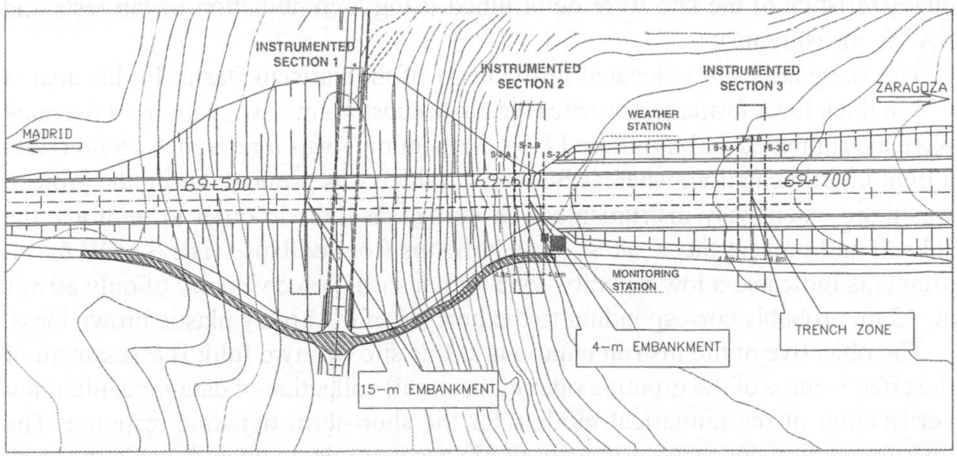

Figure 1. Location of three instrumented sections at Guadalajara test site.

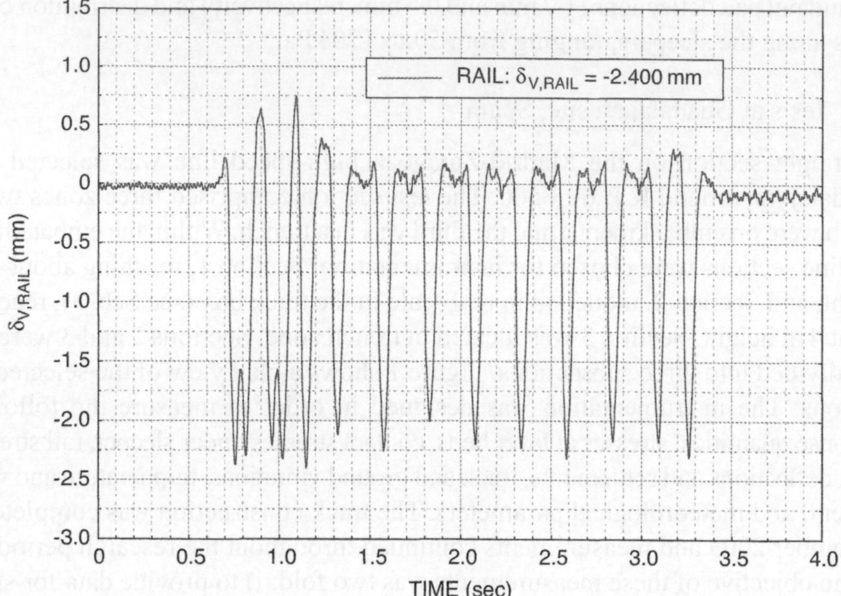

Figure 2. Absolute displacements of rail by laser measurement at Section 2.

of the order of 3.5 mm. This difference is believed to be largely due to the consolidation and increased stiffness of the track after one year of operation.

2.3 Tests at Beugnâtre, France

Beugnâtre is located in the region Nord-Pas-de-Calais, on the North Europe high speed line between Paris and Lille at 140 km North of Paris. The geotechnical characteristics of the site were determined using a combination of lab tests and SASW measurements.

The Beugnâtre site is located in the heart of the Parisian Basin. In this area, a 5–8 m thick loess formation overlies the chalk substratum. Two sections of the track were instrumented and monitored (Fig. 3). Section 1 was a reference section with a light instrumentation, whereas Section 2, which was grouted during the project, was more extensively instrumented. Grouting was done in order to reinforce the subsoil and prevent the weakest loess horizons from collapse. The SASW measurements indicated a low velocity zone with a shear wave velocity of only 80 m/s at −7 m, probably corresponding to the horizon of the highly plastic brown loess.

The objective of the instrumentations at this site was two fold: i) assessment of the effectiveness of the grouting operation, and ii) collection of data for calibration/verification of the numerical models for the short-term dynamic response. The instrumentation consisted primarily of 3D accelerometers on and inside the track and on the ground surface. In addition, strain gages were installed on the rails to

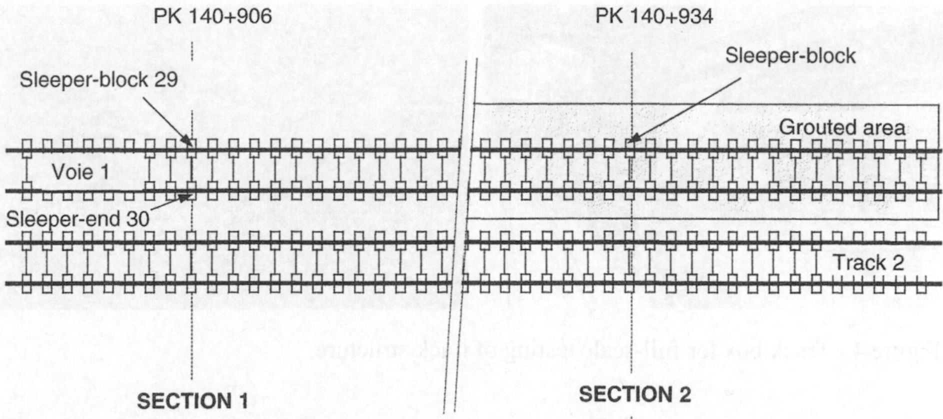

Figure 3. Test tracks at Beugnâtre.

compute the wheel loads. The underground sensors were placed inside a borehole, sealed with cement and loaded by sand. Boreholes were then filled up with clay pellets. The depths of the 12 underground accelerometers varied between 1.0 to 2.5 m from the top of the track. Sleeper accelerations were measured in the three directions.

Comparison between the measurements before and after grouting indicated a considerable reduction (in the range 20–30%) of the sleeper and embankment accelerations. In addition, altimetric measurements were performed at regular intervals to monitor the track settlements and compare the settlement rate before and after grouting. A 90 m section of track 1 was thus equipped with reference pins sealed onto the sleepers. During the 7 months period before grouting, a global average settlement of 0.72 mm was recorded. The grouting works induced an average 2 mm uplift of the track. Measurements after grouting indicated a settlement of 1.6 mm (i.e. higher than the pre-grouting settlement rate) between August and December 2004; however, no significant settlement was recorded during the first six months of 2005.

3 FULL-SCALE TESTING OF TRACK

A major achievement of SUPERTRACK was the construction of a full-scale track box testing facility by ADIF and CEDEX on the CEDEX's premises. The track box is a steel box with dimensions 4 × 5 × 21 m. It was filled with the soil from the Spanish test site and was capped by the track structure. A number of pressure cells, displacement transducers, accelerometers and geophones were installed inside the soil and on the various elements of the track. The train load was simulated by out of phase loads at 6 points on the rails by hydraulic actuators. The track box was used to reproduce the soil/track conditions at Section 2 of the Guadalajara test site (Sec. 2). Figure 4 show a photo together with a schematic representation of the track box.

Figure 4. Track box for full-scale testing of track structure.

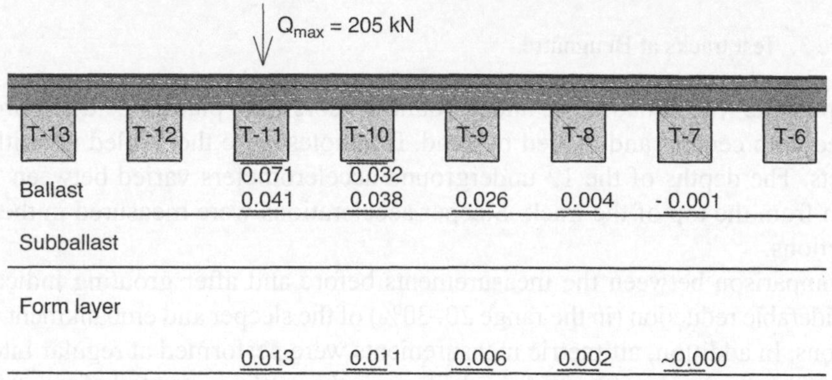

Figure 5. Measured pressures under static load in track box. Unit of pressures is MPa.

Figure 5 presents a sample data from the track box which displays the pressures measured at different locations under and along the rail under a static (axle) load of 205 kN. The figure shows the pressures measured under the sleepers, at the ballast/sub-ballast interface and at the bottom of the form layer.

Figure 6 displays the evolution of the load-displacement curves under Actuator 3 for different numbers of train passage. The figure demonstrates the gradual increase in the irreversible deformations of the track under repeated loading.

4 CONSTITUTIVE MODELING OF TRACK MATERIAL

4.1 Tests on large-scale samples

Large-scale tri-axial tests were performed in order to obtain the data required for the constitutive modelling of ballast and sub-ballast. The objective was to represent the long-term cyclic loading of the material under train traffic. The tests were performed using the vacuum-triaxial apparatus at NGI. A vacuum tri-axial works on

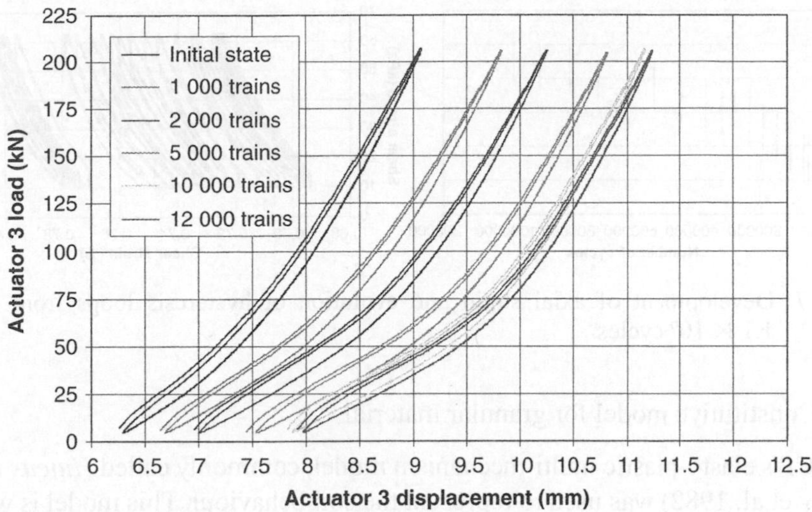

Figure 6. Load displacement curves under actuator representing different numbers of train passages.

the principle that a controlled vacuum is applied internally to the sample that is confined in the membrane; the atmospheric pressure in the laboratory thereby supplies the confining pressure. The sample size in this apparatus is 62.5 cm in diameter and 125 cm in height, thus allowing testing of rather coarse-grained materials. Four tests were performed on materials similar to those used at the French test section at Beugnâtre (Sec. 2.3). The ballast material used in Test 1 and Test 2 was 25 mm/80 mm diorite gravel from the French quarry Roy. The sub-ballast material used in Test 3 was 0/31.5 mm quartzite sandstone well graded gravel from the French quarry Vignats.

The following instrumentation was used: Load cell transducer on the piston of the MTS actuator, LVDT that measures the displacement of the piston in the MTS actuator, Vertical displacement transducer on the left side of the sample, Vertical displacement transducer on the right side of the sample, Circumferential displacement transducer at mid-height of sample, Transducer that measures vacuum relative to atmosphere. The load program consisted of several stages of cyclic loading followed by a monotonic loading to failure. The cyclic stages were performed in force control and at a cyclic frequency of 1 Hz. The monotonic stages were performed in displacement control at a displacement rate of 1 mm/min or a sample axial strain rate of 0.08%/min.

Figure 7 presents a sample of the results. The figure shows the development/accumulation of the axial strain in the sample with the number of cycles (left figure) and the evolution of the cyclic stress-strain curves during the last 100,000 cycles of the total 1.1 million cycles of loading.

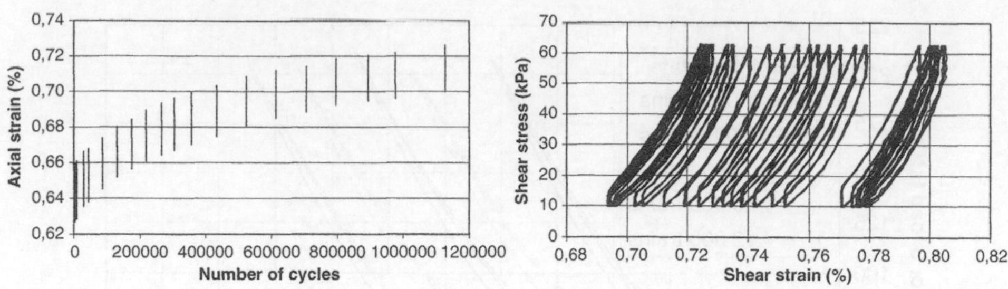

Figure 7. Development of axial strain and evolution of hysteresis loops from 10^6 to 1.1×10^6 cycles.

4.2 Constitutive model for granular material

The ECP's elasto-plastic multi-mechanism model, commonly called *Hujeux* model (Aubry et al. 1982) was used to represent the soil behaviour. This model is written in terms of effective stress and can take into account the soil behaviour in a large range of deformations. The representation of all irreversible phenomena is made by four coupled elementary plastic mechanisms; namely, three plane-strain deviatoric plastic deformation mechanisms in three orthogonal planes and an isotropic one. The model uses a Coulomb type failure criterion and the critical state concept. The evolution of hardening is based on the plastic strain (deviatoric and volumetric strain for the deviatoric mechanisms and volumetric strain for the isotropic one). To take into account the cyclic behaviour a kinematical hardening based on the state variables at the last load reversal is used.

The model is written in the framework of the incremental plasticity, which assumes the decomposition of the total strain increment in elastic and plastic parts. The elastic part is supposed to obey a non-linear elasticity behaviour, where the bulk (K) and the shear (G) moduli are functions of the mean effective stress (p′). Through appropriate choice of parameters one can control the form of the yield surface in the (p′, q) plane to vary from a Coulomb type surface to a Cam-Clay type one. Furthermore, an internal variable r_k, called degree of mobilized friction, is associated with the plastic deviatoric strain. This variable introduces the effect of shear hardening of the soil and permits the decomposition of the behaviour domain into pseudo-elastic, hysteretic and mobilized domains. Finally, an associated flow rule in the deviatoric plane (k) is assumed, and the Roscoe's dilatancy law is used to obtain the increment of the volumetric plastic strain of each deviatoric mechanism.

Calibration of the model parameters was based on the preceding tri-axial tests on ballast. Figure 8 displays the response of the model during the initial cycles of loading compared to the experimental results. The figure shows that the magnitude of simulated strain variation during one cycle is comparable to the experimental observations and they have similar form; however, the volumetric and axial strains are slightly underestimated.

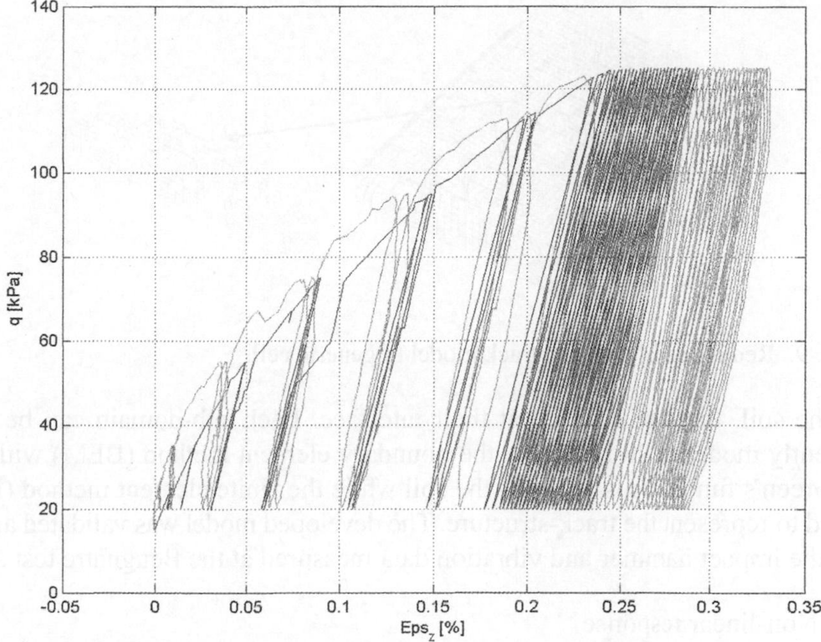

Figure 8. Comparison between simulated and experimental results in tri-axial test on ballast.

The above model was used to simulate the irreversible deformations of the ballast in the track box (Sec. 3) during the passage of a moving load corresponding to one axle during 1 million cycles. The results were found to be in fairly good agreement with the experimental data in the track box.

5 NUMERICAL MODELING OF TRACK BEHAVIOR

5.1 Linear dynamic response

An efficient, rigorous numerical model was developed at ECP for simulation of the linear dynamic response of the track-ground system (Chebli et al. 2004) . The developed model takes advantage of the spatial periodicity of the track-soil system. The periodic formulation, which was implemented by using the solution advanced by Floquet (1883), allows one to reduce the analysis of the overall system to that of a generic cell. The principle of this method is shown schematically in Figure 9.

For the dynamic analysis of the generic cell the ECP-developed computer code MISS3D (Clouteau 2000) was used. This software is based on a domain decomposition method whereby the three-dimensional domain considered (the generic cell in the present case) is decomposed into two sub-domains, the track-structure

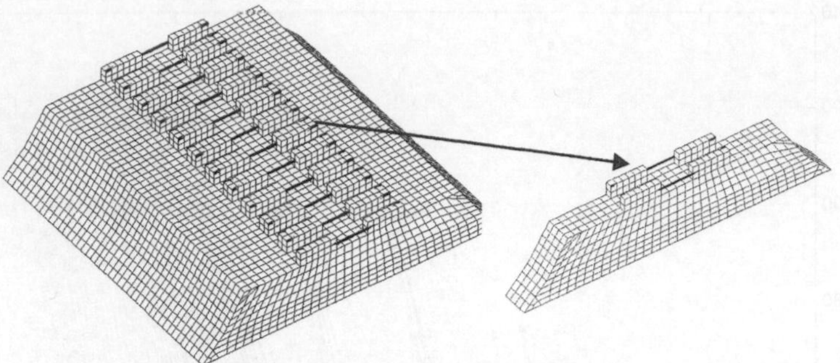

Figure 9. Reduction of periodic track model to generic cell.

and the soil, that are coupled at their interface. Each sub-domain can be independently modeled. In MISS3D, the boundary element method (BEM) with special Green's functions is used for the soil while the finite element method (FEM) is used to represent the track-structure. The developed model was validated against both the impact hammer and vibration data measured at the Beugnâtre test site.

5.2 Non-linear response

In the railway engineering community it is believed that a major cause of track settlement and deterioration is the presence of unsupported "hanging" sleepers. In addition, non-homogeneity of the track, in the form of abrupt variation of stiffness, is considered an undesirable feature in the track design. Examples of track stiffness variation can be often observed at the transition between embankments and bridges. Both these issues are believed to contribute to the increased load on the track, hence its deterioration. Therefore, an attempt was made in this study to investigate the impacts of hanging sleepers and track stiffness variation on response of ballasted tracks. For this purpose the numerical model LS-DYNA (Hallquist 1998) was used. LS-DYNA is a general-purpose finite element code for analysing large deformation dynamic responses of structures, including structures coupled to fluids. The solution methodology is based on explicit time integration. The contact force is calculated by a penalty method (contacts appear between wheel and rail and between sleeper and ballast). The code is equipped with non-reflecting boundaries.

Before employing LS-DYNA for simulation of the non-linear track response, the code was validated against the track measurements at Guadalajara test site (Sec. 2.2). A double track finite element model of the site was used in the verifications. For the non-linear analyses, a single track model was used. Figure 10 shows half of the model due to symmetry. The model consists of a wheelset, rail, rail pads, sleepers, and ballast bed. The ballast bed was modelled as a continuum with elastic/elastic-plastic material properties. The model consisted of 30 sleepers. A gap of 0.5 mm or 1 mm was introduced between the sleeper 15 and the ballast, and the increase in the dynamic forces due to this gap was investigated. The wheelset loading

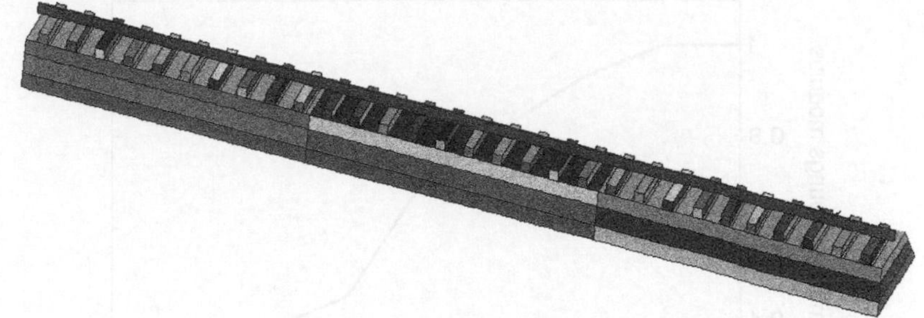

Figure 10. Track model with ballast bed of elastic and elasto-plastic material.

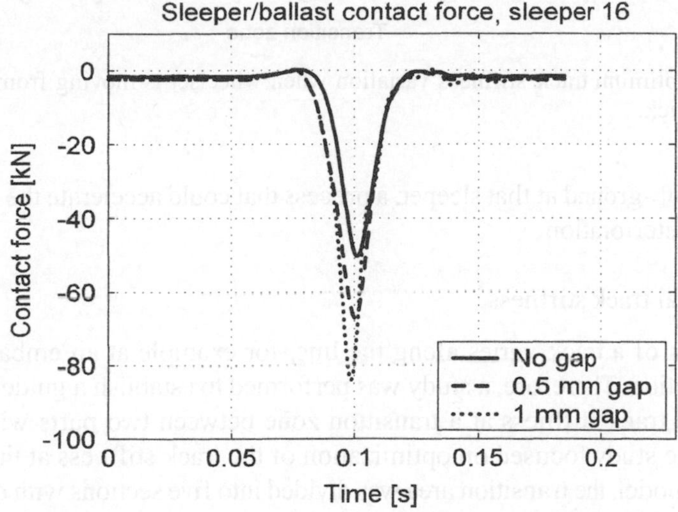

Figure 11. Sleeper-ballast contact force at sleeper 16 when sleeper 15 is hanging.

the track was moved at a speed of 90 m/s from the left (sleeper 1) to the right (sleeper 30).

The simulations showed that the sleeper/ballast contact force at sleeper 14, when sleeper 15 is hanging, increases by 20 percent due to the hanging sleeper 15. The size of the gap, 0.5 mm versus 1 mm, has no practical effect on the computed results, except that at sleeper 15 there is some contact between the sleeper and the ballast if the gap is 0.5 mm, whereas there is no contact if the gap is 1 mm. Sleeper 16 appears to be the most loaded sleeper when sleeper 15 is hanging. Figure 11 shows the computed sleeper/ballast contact force at sleeper 16. It can be seen that the sleeper/ballast contact force at this sleeper increases remarkably at a vehicle speed of 90 m/s. The figure shows that the sleeper/ballast contact force increases by 70 percent when sleeper 15 is hanging with a gap of 1 mm. Such a large overload of the ballast bed might could potentially result in non-elastic deformations of the

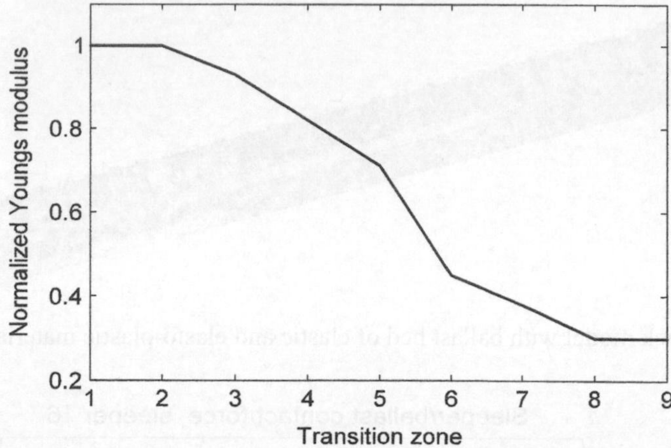

Figure 12. Optimum track stiffness variation when wheelset is moving from stiff to soft track.

ballast and sub-ground at that sleeper, a process that could accelerate the track settlements and deterioration.

5.3 Optimal track stiffness

The stiffness of a track varies along the line, for example at an embankment-to-bridge transition. Therefore, a study was performed to establish a guideline on how to select the track stiffness at a transition zone between two parts with different stiffness. The study focused on optimization of the track stiffness at the transition area. In the model, the transition area was divided into five sections with one stiffness value to be optimized in each section.

The wheel-rail contact force was selected as the objective of the optimization. The force variation (i.e. irregularity of the force) should be as small as possible when the wheel passes the transition area. Optimal track stiffness variations in the transition area, as obtained from the optimization procedure, are shown in Figure 12. The track stiffness at one end of the track model was 90 kN/mm and at the other end the stiffness was 45 kN/mm (different stiffness values were obtained by changing the modulus of elasticity of the ballast material). As expected, a smooth variation of track stiffness leads to a less dynamic loading of the track.

6 TRACK RETROFITTING

An innovative track retrofitting technique, based on the principle of hydraulic fracturing by grouting, was developed in this study (Cuellar & Vadillo 2003). One of the advantages of the proposed method is that, unlike most other soil improvement methods, it can be implemented without interruption of the train traffic. The retrofitting work was carried out on the southern embankment of the viaduct over the Ebro River on the conventional line Valencia-Barcelona. The embankment is located in Amposta, at the K.P. 180 of the above line. It was essential to carry out the soil improvement from a working platform disconnected from the railway line. Therefore, a berm was built on the side of the embankment, and the grouting was carried out without interference to the railway traffic.

The following summarises the elements/advantages of this technique: 1) possibility of improving a predefined volume of embankment, by means of fans of sleeve tubes, installed from the working platform and oriented in such a way as to cover the grouting of that volume, 2) application, at each grouting sleeve, of the necessary number of passes of treatment until the required grouting pressure is reached, 3) use, at each grouting pass, of a grouting mixture of necessary characteristics, including viscosity and setting time, placed at the convenient rate and volume, in order not to move the rails in excess of the comfortable circulation deformation limit, 4) a safe criterion to correlate the grouting pressure and the final parameters of shear strength in the treated soil so that the process of grouting could be properly controlled.

To provide the most effective treatment, the 20 m length of the embankment at the bridge approach was divided into four zones; the treatment was then applied to different depths in these zones so as to ensure a smooth variation of the track stiffness. The improvement in the embankment condition was verified using two different techniques: 1) direct comparison of physical parameters of the embankment before and after treatment; 2) indirect control by measuring the dynamic response of the track to daily traffic. PS-logging and cross-hole techniques were used to evaluate the variation with depth of the shear-wave velocity in the embankment before and after grouting. The data indicated a noticeable increase in the embankment stiffness of the order of 2.5 (considering that stiffness is proportional to square of shear wave velocity). In addition, a number of measurements were made of the track deformations under train traffic before and after the grouting. The ratio between the maximum deformations in these measurements was 1.26. In assessing the effectiveness of grouting, however, one should take into consideration that the treatment is basically applied to the embankment without any effect on the track structure (i.e. ballast and sub-ballast). By isolating the stiffening effect of the embankment it was demonstrated that the modulus of deformation of the grouted embankment increased to about 2.5 times the initial modulus.

ACKNOWLEDGEMENT

This research was performed under Contract G1RD-CT-2002-00777 in the Competitive and Sustainable Growth Programme of the European Commission's 5th Framework Program. The authors would like to express their appreciation to the European Commission for the partial financial support to this research. The authors would also like to thank the other members of the research group: V. Cuéllar (CEDEX), L. Schmitt (SNCF), A. Lozano (ADIF), T. Dahlberg (LU), A. Modaressi & H. Chebli (ECP), A. Pecker & F. Ropers (GDS), E. Berggren & E.L. Olsson (BV) and R. Dyvik (NGI) for their contributions to this paper.

REFERENCES

Aubry D., Hujeux J.C., Lassoudire F., and Meimon Y. 1982. A double memory model with multiple mechanisms for cyclic soil behaviour. *Int. Symp. Num. Mod. Geomech*, pp. 3–13.

Chebli, H., Clouteau, D. and Modaressi, A. 2004. Three-dimensional periodic model for the simulation of vibration induced by high-speed trains. *Italian Geotech. Review*.

Clouteau D. 2000. ProMiss 0.2, Ecole Centrale de Paris, Châtenay Malabry, France.

Cuellar, V. & Vadillo, S. 2003. Instrumentation of a high-speed track in Spain for the European *Supertrack. Proc. 3rd World Cong. on Railway Research*, Edinburgh, Scotland, Sept. 29 – Oct. 3.

Floquet, M. G. 1883. Sur les équations différentielles linéaires à coefficients périodiques. *Annales de l'Ecole Normale*, vol. 12.

Giraud, H. & Schmitt, L. 2004. Experimental Site on high speed line in Northern France for European Project SUPERTRACK. *Proc. 7th International Railway Engineering Conference & Exhibition*, London, UK.

Hallquist, J.O. 1998. *LS-DYNA Theoretical Manual*, Livermore Software Technology Corporation, 2876 Waverly Way, Livermore, California.

Kaynia, A.M. 2006. Sustained performance of railway tracks. Final report of SUPERTRACK to European Commission.

Lundqvist A. & Dahlberg T. 2003. Dynamic train/track interaction including model for track settlement evolvement, *Proc. 18th IAVSD Symposium*, Atsugi, Japan, August 24–29.

Lundqvist, A. & Dahlberg, T. 2005. Railway track stiffness variations – consequences and countermeasures. *Proc. 19th IAVSD Symposium on Dynamics of Vehicles on Roads and Track*, Milan, Italy, Aug 29 – Sept 2.

3

Frost Design Method for Roadways and Railways – State of the Art in France

C. Mauduit & J. Livet
Bridge and Roadway Regional Laboratory, French Equipment Department, Nancy, France

J-M. Piau
Bridge and Roadway Central Laboratory, French Equipment Department, Nantes, France

ABSTRACT: Frost design method has in France a major influence on pavement design and therefore on the cost of road infrastructures. In particular in the North East of the territory, structures, designed versus mechanical stress, have to be "overdesigned" according to thermal ones. This paper presents the state of the art French frost design approach for roadways and current research projects carried out to improve this method. It also deals with the railway problematic : the old methodology always used up to now and its project of modernization, based on an adaptation of the new road methodology developed by the network of Laboratoires des Ponts et Chaussées.

1 INTRODUCTION

This paper deals with the French state of the art frost design method for roadways and railways. It presents in the first part, after a brief introduction on the pavement rational approach for roadways, the five steps to follow to design a road taking into account the risks of damage in frost and thaw periods. Then it focuses briefly on the different ways of investigation to optimize current pavement design. In the second part, the methodology employed for a long time for railways is presented. And finally, the current research project on the revision of the frost design method for railways and in particular for high speed lines, inspired from road design, is detailed.

2 ROAD INFRASTRUCTURES

The pavement design method consists in a rational approach based on the knowledge of the mechanical characteristics of the materials employed (normative stages), their manufacturing processes (control) and implementation. It allows adjustment

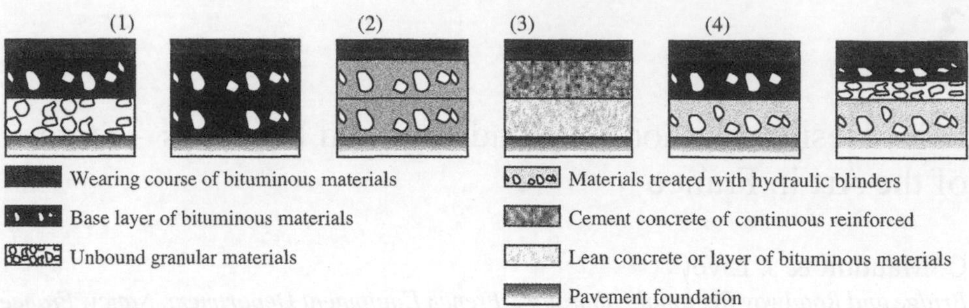

Figure 1. The six families of pavement structure.

of the thickness of the structures to the local context of bearing capacity of the roadbed and of traffic, according to the materials used and the maintenance policy adopted. Six families of structures are encountered in France : flexible pavements (1), thick bituminous pavements (2), pavements with base layers treated with hydraulic blinders (3), rigid pavements (4), composite (5) and inverted pavements (6).

The frost design method consists in a "verification" with respect to freeze/thaw phenomena, making sure that the roadway design as determined from mechanical calculations can withstand, without notable damage, a given winter chosen as a reference.

With the exception of very large construction projects or special cases, pavement design in France is not carried out case by case. Each road owner has published a document which describes its policy and offers a number of recalculated mechanical and thermal solutions for its network (CFTR, 2003) (DRCR, LCPC, SETRA, 1998), (LCPC-SETRA, 1994 & 1997).

2.1 Roadways frost design method (LCPC-SETRA, 1994 & 1997)

Verification of the frost/thaw behavior involves comparing:

- IR, the atmospheric frost index chosen as reference, that characterizes the severity of winter, according to the area, against which we choose to protect the pavement. The winter chosen as the reference depends on road manager policy. Two reference indexes are usually retained : the exceptional winter frost index (the highest index since 1951) and the non exceptional rigorous winter (winter with 10 years occurrence).
- and IA, the atmospheric allowed frost index that the pavement is able to withstand. This index is evaluated according to the frost susceptibility of the subgrade, the thermal protection and mechanical function fulfilled by the pavement.

The pavement structure is designed so that the allowed frost index IA of the pavement is higher than the reference frost index IR. In the opposite case, the pavement has to be modified (reduction of the frost susceptibility of materials or

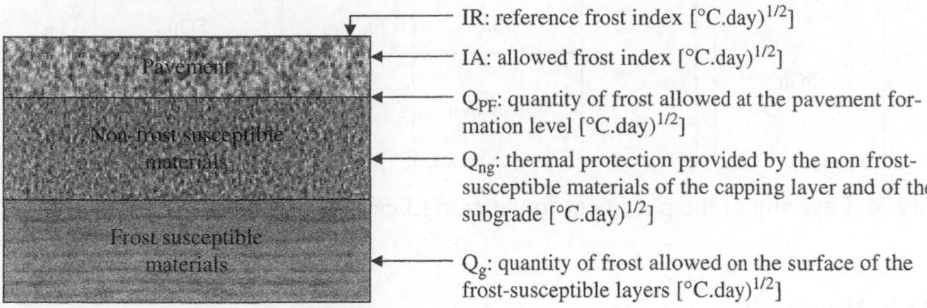

Figure 2. Frost verification principle.

increase in the thickness of the non-frost susceptible layers) until a positive frost testing is reached. IR can also be larger than IA if we accept that thaw barriers will be installed to protect the pavement during thaw periods.

The following method, in five steps, is used to determine the allowed frost index for the pavement, working from the subgrade up through to the pavement surface.

2.1.1 *Step 1: Examination of the frost susceptibility of the pavement foundation*

2.1.1.1 Frost-susceptibility of the materials

Depending on their nature, the soils and aggregates are sensitive in varying degrees to the phenomenon of frost intake or "cryosuction" that can be assessed in a laboratory test, the swelling test (NF P 98-234-2). The value of the slope of the swelling curve determines the class of frost susceptibility:

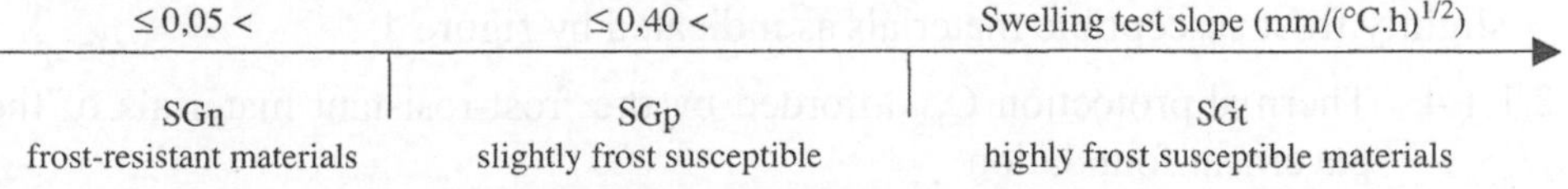

2.1.1.2 Layering of the pavement foundation

The pavement foundation (subgrade and capping layer) is layered so that frost susceptibility increases with depth. Thus, if a layer of materials is located under a more frost-susceptible layer, we treat it as if it has the same frost susceptibility as the layer above. This leads to three configurations (figure 3), for which we make a distinction in this way between the slightly frost susceptible or highly frost susceptible materials and the top layers (possibly of zero thickness for resistant materials hn) that are frost-resistant.

2.1.1.3 Allowed quantity of frost Q_g transferred to the frost-susceptible materials of the subgrade

Where the subgrade includes frost-sensitive layers, the quantity of frost Q_g allowed at the surface of these frost susceptible materials is determined as follows:

In case (a), there is no problem of frost verification.

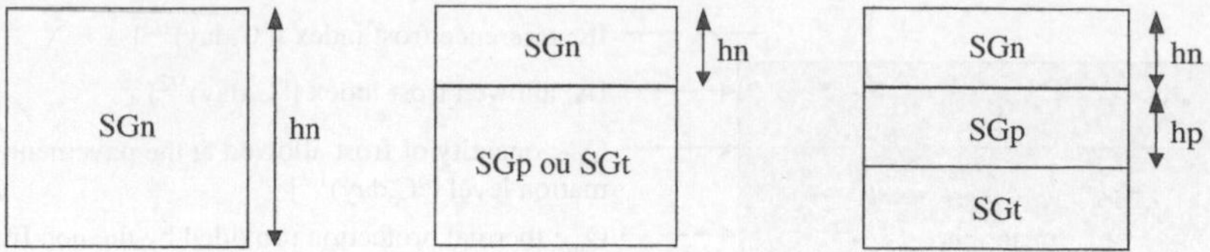

Figure 3. Layering of the pavement foundation (3 cases).

Table 1. Values of Q_g.

p (mm/(°C.hour)$^{1/2}$)	≤0.05	0.05<p≤0.25	0.25<p≤1	p>1
Q_g ((°C.day)$^{1/2}$)	∞	4	1/p	0

Table 2. Values of the coefficients A_n.

Materials	A[1]	B[1] et C[1]	D[1]	LTCC[2]	CV[3] SC[4],SL[5]
A_n (°C.days)$^{1/2}$/cm	0.15	0.13	0.12	0.14	0.17

[1] soil classes defined by standard (French standard : NF P 11-300)
[2] lime cement-treated silt, [3] fly-ash, [4] cement-stabilized sand, [5] slag-stabilized sand.

In case (b) the allowed quantity of frost is fixed according to the frost susceptibility of the material (p: slope obtained by the swelling test [mm/(°C.hour)$^{1/2}$]) as:
In case (c) the quantity of frost allowed Q_g depends on the thickness (hp) of slightly frost susceptible materials as indicated by figure 4:

2.1.1.4 Thermal protection Q_{ng} afforded by the frost-resistant materials of the pavement foundation

The thermal protection afforded by the frost-resistant materials of the pavement foundation depends on both their nature and thickness.

The value of Q_{ng} [°C.day)$^{1/2}$] is given by the formula: $Q_{ng} = A_n.hn^2/(hn + 10)$

with, hn: thickness of the frost-resistant layer, in cm,
An: coefficient depending on the nature of the capping layer material (see table 2).

2.1.2 *Step 2: Mechanical analysis*

The reduced bearing capacity of the frost-susceptible materials of the subgrade on thawing due to the increased water content leads to greater stress in the pavement layers than in normal periods.

- With thick pavements (bound layers of a total thickness exceeding 20 cm), a degree of frost penetration into the frost-susceptible layers of the subgrade is acceptable, if the resulting increase in stress during thaw periods can be limited. Therefore we accept the transmission into the underlying pavement formation level of a quantity of frost, termed Q_M, in addition to the terms Q_g and Q_{ng} defined earlier.

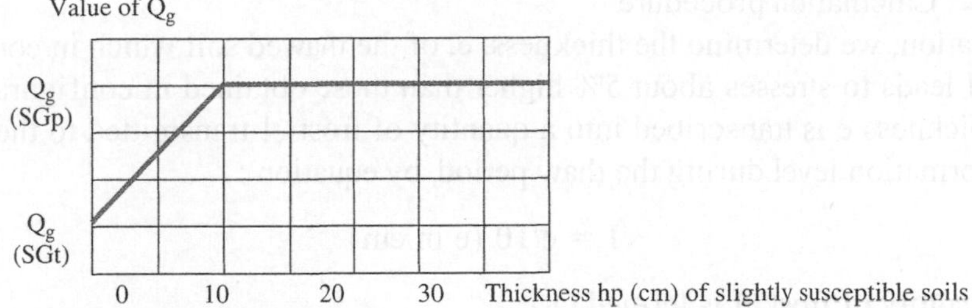

Figure.4. Quantity of frost allowed Q_g on the surface of slightly frost-susceptible layers.

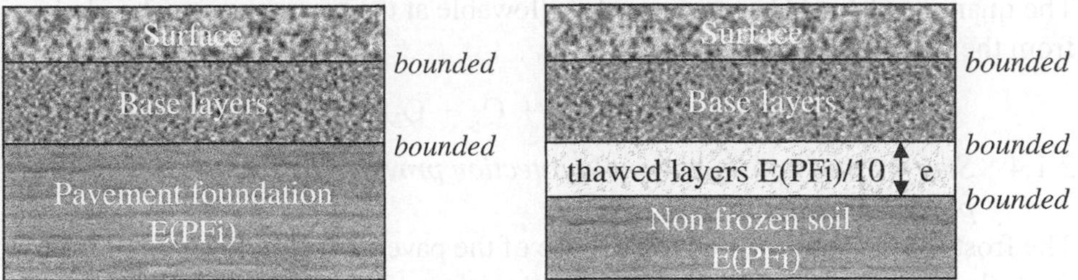

Figure 5. Adopted configuration for the mechanical phase of the verification.

This quantity Q_M is determined according to the method outlined below.

– For thinner pavements (bound layers less than 20 cm thick) no mechanical verification will be done; Q_M will be taken as nil.

2.1.2.1 Principle

The mechanical behavior of the structure during the thaw period is checked by limiting the increased stress in the pavement structure to about 5% over the normal situation other than the frost/thaw season. During the thawing phase, calculation realized with Alizé-LCPC software (multi-layer linear elastic analysis), is based on the following hypotheses:

– the modulus of the thawed subgrade layer is taken as equal to E(PFi)/10;
– the modulus of the remainded subgrade is that of the subgrade outside frost/thaw period,
– the top and bottom interfaces of the thawed layers are bonded;
– the mechanical characteristics of the layer of thawed soil during the thaw period are taken as constant throughout the duration of thawing.

All of the other data involved in the problem (modulus values, interface conditions) comply with the usual values adopted for design calculations.

2.1.2.2 Modelling the structure

The structure is defined according to two configurations (figure 5):

– the first is that of the structure in normal operation, on non-frosted soil (configuration I);
– the second corresponds to the presence of a layer of thawed soil (configuration II).

2.1.2.3 Calculation procedure

By iteration, we determine the thickness, e, of the thawed soil which in configuration II leads to stresses about 5% higher than those obtained in configuration I. This thickness e is transcribed into a quantity of frost $\sqrt{I}$ transmitted to the pavement formation level during the thaw period, by equation :

$$\sqrt{I} = e/10 \text{ (e in cm)} \tag{1}$$

This quantity of frost $\sqrt{I}$ is termed Q_M.

2.1.3 *Step 3: Quantity of frost allowed at the formation level*

The quantity of frost Q_{PF} we consider allowable at the formation level is deduced from the preceding terms:

$$Q_{PF} = Q_{ng} + Q_g + Q_M \tag{2}$$

2.1.4 *Step 4 : Study of the thermal protection provided by the pavement structure*

The frost index It transmitted to the base of the pavement structure is derived here from the frost index IS at the pavement surface. Two approaches are possible, depending on the degree of accuracy sought in the analysis: a thermal calculation of the propagation of frost in the pavement or the use of simplified equations.

2.1.4.1 Calculation of frost propagation through the pavement

Frost propagation through the pavement is studied using a digital computation software (for instance the finite differences computation software GEL1D developed by the LCPC (Caniard et al., 1975) based on Fourier's model. To use this type of software requires:

- a description of the structural geometry and thermal characteristics of each of its layers,
- a definition of boundary conditions for temperature: the initial temperature conditions imposed upon the pavement are defined by the following profile: the temperature is 1°C on the pavement surface and varies in linear form up to 14°C at 10 m depth under the pavement formation level, we leave it at 14°C below this depth.
- the law of evolution of the surface temperature: the surface temperature evolves over time according to a hyperbolic law with an initial slope of −0.833°C/hr.

The thermal calculation enables determining the relation between the surface frost index, termed IS, and the frost index It transmitted to the base of the pavement structure.

2.1.4.2 Simplified method

Based on calculations of the preceding type, results are simplified by linearizing the relation between $\sqrt{IS}$ and $\sqrt{It}$. For a homogeneous pavement structure of thickness h, we would allow an expression as follows, with a and b coefficients depend on the nature of the materials:

$$\sqrt{IS} = (1 + ah)\,\sqrt{It} + bh \qquad (3)$$

2.1.5 *Step 5: Determination of the allowed atmospheric frost index IA*

2.1.5.1 Relation between the surface frost index and the atmospheric frost index

At average altitude, with little to average sunshine and an atmospheric frost index not exceeding 210°C.day, the convection and radiation phenomena at the pavement surface are accounted for by the equation:

$$IS = 0.7\,(IAtm - 10) \qquad (4)$$

with IS: frost index at the pavement surface (°C.day);
IAtm : atmospheric frost index (°C.day).

For all the other cases (very severe frost, much sunshine), an individual study will be needed.

2.1.5.2 Allowed frost index IA

The quantity of frost Q_{PF} that can be allowed at the pavement formation level determines the frost index (It) that can be transmitted to the base of the pavement:

$$Q_{PF} = Q_{ng} + Q_g + Q_M = \sqrt{It} \qquad (5)$$

Based on the thermal calculation or the approximated relation between $\sqrt{IS}$ and $\sqrt{It}$ given previously, we determine the value of IS associated with Q_{PF}.

The atmospheric frost index IA, corresponding to IS can therefore be deduced :

$$IA = IS/0.7 + 10 \qquad (6)$$

2.2 Current research

Current research has two principal objectives:

- Validate pavement frost design method which is sometimes based on "arbitrary" assumptions, even if the method has never really be refuted up to now.
- Optimise the method, which presents large safety coefficients, in order to have a better control of uncertainty intervals of all the parameters employed.

Current research focuses on four principal topics:

2.2.1 *Reference frost indexes notion*

- Improvement of the accuracy of the definition of the reference frost index by an establishment of a Geographic Information System taking into account more meteorological stations.

- Exploratory works on the risk notion taken by the road owner and on the choice of new more pertinent reference indexes.

2.2.2 *Frost behaviour of soils and materials*

- Statistical analysis of the national database of frost heaving tests (more than 1600 soils and granular materials treated or not) according to the French standard NF P 98-234.2, in the aim of characterising frost susceptibility by geotechnical or mechanical indicators.
- Study of the hydraulic blinders taken below first frost conditions.

2.2.3 *Thermal modelling*

- Validation of thermal models GEL1D and CESAR-GELS: study on the variability of the models' input parameters, adjustment of simulated frost front on temperature monitored on test beds in order to highlight limits of the models and modifications to provide in order to improve frost penetration simulations (taking into account cryosuction for instance, etc.). Several experimental pavements have been monitored for this purpose.
- Laboratory tests on the thermal conductivity of materials employed in pavement structures to increase knowledge on the thermal transfer in base, foundation and capping-layers.

2.2.4 *Mechanical approach*

- Improvement in fatigue damage and diminished load-bearing capacity under the combined effect of traffic and frost-thaw cycles: a project developed with the Ministry of Transportation of Quebec on the behavior of roadways during severe frost conditions aims to calibrate the damage models to the pavement deterioration observed on test beds, comparing insulated and uninsulated pavements.

3 RAIL INFRASTRUCTURES

3.1 Current frost design method

The Société Nationale des Chemins de Fer français (SNCF) reference document for track beds layers design takes into account frost protection, but with an approach based on the old road design methodology (SNCF, 1995 & 1996). This approach, based on a frost verification too, is summed up below.

3.1.1 *Frost susceptibility*

Soils are divided into three categories, according to their frost susceptibility: frost resistant (SGn), slightly frost susceptible (SGp) and highly frost susceptible (SGt) soils. A frost resistant soil is defined here as a soil where frost and thaw period do not generate unacceptable deformations for the track's levelling.

Table 3. Frost susceptibility classes adopted in the absence of frost swelling test.

Soils classification (NF P 11-300)	Frost susceptibility classes, adopted without frost swelling test
D1, D2, D3, R41, R42, R61, R62, F31, F7	SGn
B1, B2, B3, B4, R11, R43, F11, F12	SGp
A1, A2, A3, B5, B6, R12, R13	SGt

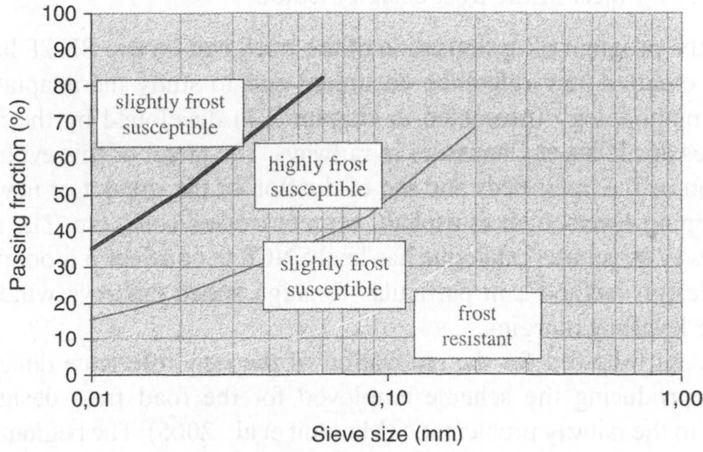

Figure 6. Abacus of definition of soils frost susceptibility.

Determination of frost susceptibility of soils in the current reference document of SNCF, can be defined, following different steps, according to the degree of accuracy sought, from:

– Indications mentioned in the table 3 presented below where determination of frost susceptibility is realized according to the French soils classification.
– Abacus of definition of frost susceptibility which is determined according to the grading curve of the 0/2 mm fraction of the soil.
– Frost swelling test. For formation-level or structure where frost and thaw can generate injurious deterioration to the safety of railway circulation, previous checks are insufficient. If frost front must penetrate layers on 5 cm, a frost swelling test, according to the French standard (NF P 98-234.2) is recommended.

3.1.2 *Frost design method*

Frost design method consists in a comparison between :

– Frost depth read on a France map, corresponding to the frost depth of a very hard winter, considered statistically with 20 years occurrence,
– And protection thickness, defined as the thickness of the frost resistant materials, recovering the first frost susceptible layer, as the latter will not be penetrated on more than 5 cm thick.

If this verification is negative, one of the frost resistant layers (ballast, sub-layer or non-frost susceptible capping layer) has to be thickened by steps of 5 cm, until protection thickness will be superior to the frost depth.

Frost risk is more or less important according to soil susceptible degree and hydrologic conditions. Therefore, frost depth can be pondered by a coefficient of 0,9 when soils are not highly frost susceptible and when hydrologic conditions are very good (correct drainage appliance).

3.2 Revision project of the frost design method

As part of the program of optimisation of the track bed layers, SNCF has recently decided to create a new reference document and to study the adaptation of the new road methodology (presented in chapter 2.1) developed by the network of Laboratoires des Ponts et Chaussées to railways. The target of this evolution is the optimisation of the track beds and the evaluation of the impact of new materials used in capping-layers such as asphalt, cement treated soils, etc. The revision of typical railway structures catalogue has lead SNCF to consider a modernization of its "frost design method", in particular for high speed railways which required very severe levelling margins.

The process followed for the realization of the new reference document consisted in reproducing the scheme employed for the road frost design method, adjusting it to the railway problematic (Mauduit et al., 2005). The beginning of work on this project is presented below.

3.2.1 *Different problematics between pavement and railway structures*

In the railway field, the problematic is different from the one for roads:

– frost issues are rather estimated through the notion of the materials' heaving during frost periods, considering the necessity of a perfect evenness of the structure,
– railway approach has to take into account high speeds, where any defect of lengthwise level would damage the structures and then reduce their lifetime,
– railway structures can not integrate thaw barriers or adjunction of material on the surface, so design method assumptions have therefore to be more severe,
– lastly, rain penetrates the structure and traffic solicitations involve ballast's attrition, that is continuously modifying physical properties of railway components.

3.2.2 *Additional investigations to carry out*

As both fields differ on a few points, some questions have been raised and need to be solved by additional tests or studies. Among them, different work has been carried out on:

3.2.2.1 Frost susceptibility of railway materials and the way of taking it into account

A statistical study of the LRPC national database of frost swelling tests (1600 soils and granular materials treated or not) has been achieved to validate ST590B criteria versus frost for sub-layers and capping layers. This work has shown that ST590B specifications are not severe enough to guarantee the use of frost resistant soils.

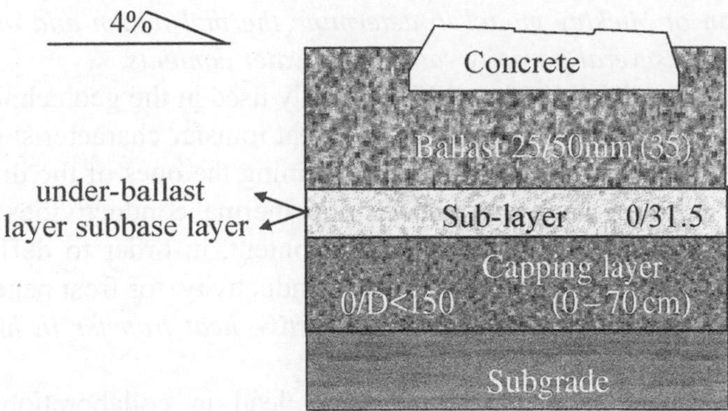

Figure 7. Typical structure for new railways.

Moreover, interpretation of the result of the frost swelling test has been discussed. The parameter admitted for roadways is the slope of the heaving frost test which suggests a loss of bearing capacity during thaw period. For railway problematic, it seems to be absolute heaving during frost period which is the worrying parameter.

3.2.2.2 Thermal exchanges on surface structure

The relationship between air and surface temperature and between corresponding indexes established for roads (IA = IS/0.7 + 10), is not directly transposed to the railway field because radiative exchanges are very different from the ones on pavements. Relationships have been highlighted from two railway test beds, but a sole equation could not be determined. Additional measurements are planned on railway lines to determine a more realistic relationship.

3.2.2.3 Physical properties of ballast

Thermal behaviour of ballast is one of the most important parameters for the definition of the frost design method (fundamental coefficient with a significant incidence in thermal modelling). Moreover, physical properties of ballast evolve during the lifetime of the structure due to combined effects of climate and mechanical solicitations. Intrinsic characteristics are directly linked with external factors: precipitations increase the water content of the ballast (temperature acting on the phase of this water) and traffic solicitations involve ballast's attrition and rising of particles from sublayers, increasing pore filling process and modifies its specific weight. So, several investigations have been carried out to characterise ballasts :

– *Experimental measurements to characterise ballast thermal conductivity.* Different sets of experimentation on two type of ballast, at a steady state and in dry conditions, have been made (Mauduit, Livet 2003–04). Measurements of the thermal conductivity of the ballast porous medium (in its granular shape), with two levels of pore filling have been carried out and have been cross compared with measurements on rock sample (solid phase).

- *Application of Mickley model to determine thermal frozen and unfrozen conductivities for several porosity values and water contents.*
  The Mickley model (Mickley, 1951), usually used in the geotechnical field for its simplicity, allows to approximate the heat transfer characteristics of a composite medium (as a porous one) by combining the ones of the different components. It has been used here to determine thermal conductivities of clean and dirty ballasts, for their stabilised water content, in order to define the most restrictive couple water content/thermal conductivity, for frost penetration.
- *Theoretical modelling to approach convective heat transfer in highly porous materials.*
  Initial experiments and modelling work lead in collaboration with Paris University, show that the convection phenomena might be of first importance in porous materials such as ballast, a fact that has been also pointed out by other researchers (Goering, 2000), (Lai Y., 2004).

As the same modeling tool as for the road method is used (Gel1D and CESAR-GELS), work has consisted in describing the heat transfer as a conductive one assuming that the contribution of convection is described by including an isotropic increase of heat conduction (frozen risk being a monotonous function of effective conductivity). Three typical convective configurations are evaluated for open or closed embankment : one without convection, one with relatively intensive natural convection and one when a huge forced convection occurred. Different scenarii are being tested to know the list of convection and to define the majorating conductivity coefficient to retain.

3.2.3 *Follow-up of the work to establish the new reference document*

Once the conductivity values of the different ballast usable in railway lines have been defined, all the possibilities of assembling layers, in nature and thickness, will be tested. CESAR-Gels and Gel1D models will allow determination of thermal transfers in the different assembling of ballasts and sub-layers (IS = f(It)), and also thermal protection of the different capping layers (establishment of the relationships between Q_{ng} = f(hn)). Acceptable surface and atmospheric freezing index for the different assembling will be determined so as the frost front will reach the top of the frost susceptible soil, according to the hypothesis $Q_g = 0$.

This methodology, focused at first time on high speed lines, aims to be extended in the future to classic railway tracks. A dynamic tool is wished by the awarding authority in order to include easily new materials, treated ones in particular.

4 CONCLUSION

French frost design method for roads is based on thermal modelling of structures and winter characterization through the freezing index concept. This method, used for more than 25 years has never really been refuted up to now. Current research is being carried out in order to optimize the method which presents large safety coefficients.

Work engaged on revision of the frost design method for railway lines has shown that the transfer from the pavement method is delicate. The different problematics have raised some questions on which work is still being currently conducted. In particular, use of atypical materials for roads builders, such as ballast, needed additional laboratory tests in order to define the impact of ballast physical properties on frost penetration.

REFERENCES

Caniard, L., Dupas, A. Frémond, M. and Lévy, M., 1975. Comportement thermique d'une structure routière soumise à un cycle de gel-dégel, simulations expérimentales et numériques. *Vième congrès international de la Fondation Française d'Etudes Nordiques*, France.

CFTR (Comité français pour les techniques routières), 2003. *Pavement design: a rational approach*. European Roads Review, special issue RGRA 1.

DRCR, LCPC, SETRA, 1998. Catalogue des structures types de chaussées neuves – circulaire 77-1156 du 5 décembre.

Goering et al., 2000. *Convective heat transfer in railway embankment ballast*. Ground freezing 2000, communication. Rotterdam. p31–36.

Lai, Y. et al., 2004. *Adjusting temperature distribution under the south and north slopes of embankment in permafrost regions by the ripped-rock revetment*. Cold Regions S&T, 39:1.

LCPC-SETRA, French design manual for pavement structures, 1997. Translation of the December 1994 French version of the technical guide.

Mauduit, C., Livet, J., 2004. *Analyse du fichier national des essais de gonflement au gel. Caractérisation et étude de sensibilité des propriétés physiques de quelques ballasts*. Rapport de recherche LCPC.

Mauduit, C., Livet, J., Peiffer, L., Robinet, A., Lefebvre, G., 2005. *Revision of the French reference document of frost design method for High Speed Railway Lines*. 7th International Conference on Bearing Capacity of Roads, Railways and Airfields, June 27–29, Norway.

Mickley A.S., 1951. *The thermal conductivity of moist soil*. Am. Inst. Elec. Eng. Trans.

SNCF, Direction de l'infrastructure, 1995. *ST N°590 B. Spécification technique pour la fourniture des granulats utilisés pour la réalisation et l'entretien des voies ferrées*.

SNCF, Direction de l'Equipement et de l'aménagement, 1996. *Notice générale EF2C20 N°3. Dimensionnement structures d'assise pour la construction et réfection des voies ferrées*.

4

Influence of Impact Load on Base Course and Sub-grade by Circulation

Y. Shioi
Hachinohe Institute of Technology, Hachinohe, Aomori, Japan

T. Sakai
Applied Research Co. Ltd. Tsukuba, Ibaraki, Japan

ABSTRACT: Asphalt pavements are supported by a base course and a sub-grade, which are designed using a Modified CBR test and a CBR test. The durability of the pavement depends on the volume of traffic of heavy vehicles in circulation and the bearing capacity of the sub-grade. If the surface of the pavement is smooth and the base course is thick enough, the pavement will remain intact regardless of the number of heavy vehicles. However, the damage to the pavement will increase rapidly if cracks or corrugations occur. The main cause is thought to be unexpectedly large impact waves with very short periods, which break down the surface course and then proceed to collide with the weak aggregate of base course, bring excess pore water pressure in the sub-grade. However, the measurements taken to make this process clear using a special accelerator for high cycle waves produced some unexpected results.

1 INTRODUCTION

In constructing a durable pavement the bearing capacity of the base course and the sub-grade is very important, although the surface course should also be of the required strength. Figure 1 shows the standard composition of the asphalt pavement by Japan Road Association.

The total thickness of pavement (surface + binder + base courses) is determined mainly by the value of CBR for the sub-grade and the volume of traffic of heavy vehicles. The thickness and strength of the surface and binder courses is based on the values from the Marshall stability test and the volume of traffic of heavy vehicles. The quality of the base course is evaluated using the modified CBR test and the axial compression test.

Asphalt pavements have a flexible surface that is elastically supported by the base course. As long as the volume traffic of heavy vehicles is within the design

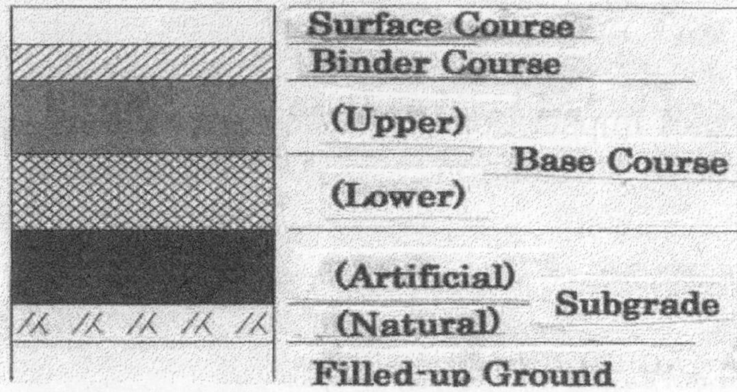

Figure 1. Standard composition of asphalt pavement.

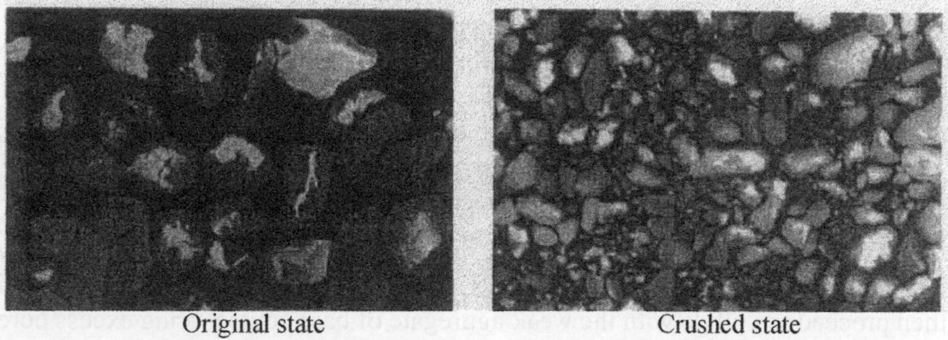

Photo 1. Comparison of aggregate in an original and a crashed state.

value, the surface should remain intact. However, if there is excess traffic volume or vehicles with excessive load, the surface course will deteriorate and fine cracks, holes and corrugation will occur. After this, vehicles will impact the pavement and the breakages in the surface will progress rapidly. In particular, deterioration in the base course will severely affect the surface.

We know from daily experience that when a heavy vehicle passes over a hole in the surface, a strong impact generates and spreads. The impact transfers into the ground and may crush or powder weak aggregate of the base course. Photo 1 shows a comparison between aggregate in an original state and crushed aggregate in a tri-axial compression test. As shown in Photo 1, crushed or powdered aggregate no longer has much bearing capacity and deformation and creep in it steadily becomes larger.

2 MEASUREMENT OF VIBRATION INCLUDING IMPACT

To examine the influence of impact at the edge of a road, we measured the vibration due to a vehicle passing over a stepped section. The sensor used was a part of the

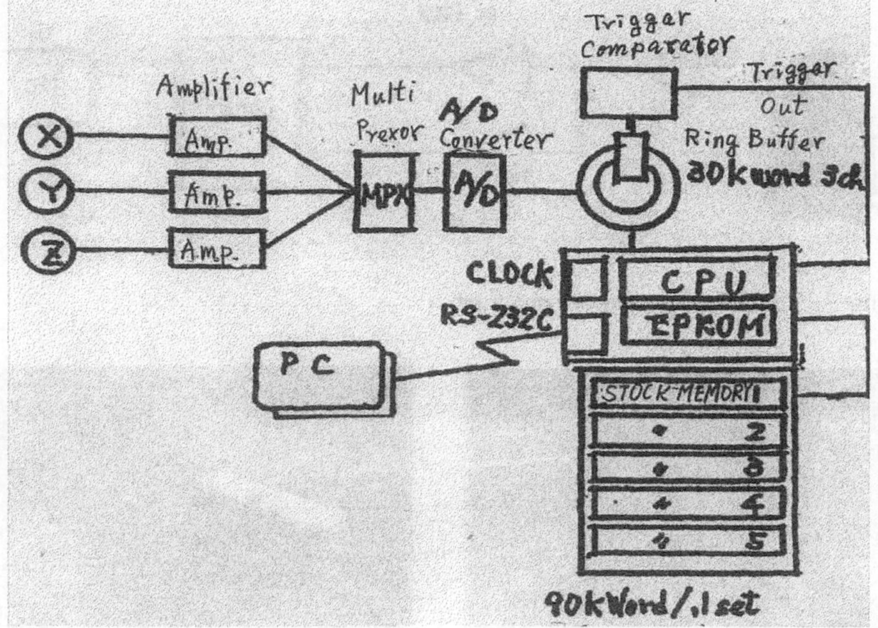

Figure 2. Diagram of seismograph for high frequency waves.

seismograph newly developed to measure high cycle seismic waves below 500 Hz. The step-up of the system is shown in Figure 2 and it has the following specifications.
Sensor: piezo-register type piezo-electric element

Number of input channels: 3 ch
Measurement range: ±5 g
Frequency range: 500 Hz
Interval of samples: 2 kHz
A/D accuracy: 12 bits
Trigger: external (horizontal vibration sensors)
Length of data: up to 300 sec
1/F: RS-232C
Records directly to a PC

The simplified one-way sensor shown in Figure 3 was taken from this system, Photos 2 and 3 show the sensor in position on the surface of a road. It has the following specifications.

Accelerometer: PCB393A03
Components of measurement: Vertical component only
Range of measurement: micro vibrometer 0.1 gal ~100 gal
wide range vibrometer 10 gal ~ 10 G
Length of time of measurement: 1 min. (maximum)/1 record

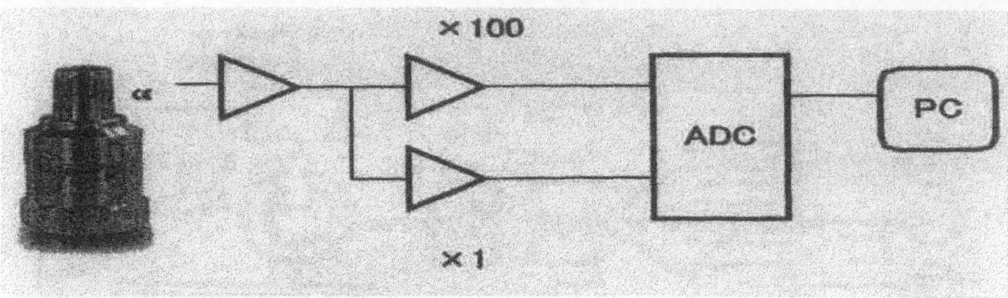

Figure 3. Simplified one-way seismograph.

Photo 2. The sensor in position.

Photo 3. Close-up of sensor.

Accuracy of AD converter: 12 bit 2ch
Trigger level: optional
Frequency of measurement: 1 kHz AD speed 2 kHz(0.5 ms)

A series of travel tests was performed on a road with a truck in the grounds of a research institute. The empty weight of the truck was 4 ton and the total laden weight was 8 ton. A board 25 mm thick, 2.5 m long and 200 mm wide was fixed on the road. The speed of the vehicle speed was set at 40 km/h.

Figures 4 and 5 show a record of the acceleration waves measured and the power spectrum of the density function of the waves for the 4 ton truck passing over the board. Figures 6 and 7 are the equivalent record and the spectrum for the same truck loaded to weigh 8 tons. The records and spectra for the 4 ton truck and the 8 ton truck are almost identical. The records for the loaded truck (8 ton) show slightly larger values than the records for the empty truck (4 ton).

3 MEASURED RESULTS AND NOTES

The fact that the shapes of the two sets of waves and spectra are identical with two peaks near 20 Hz, means that the influence of loading is small probably due to the

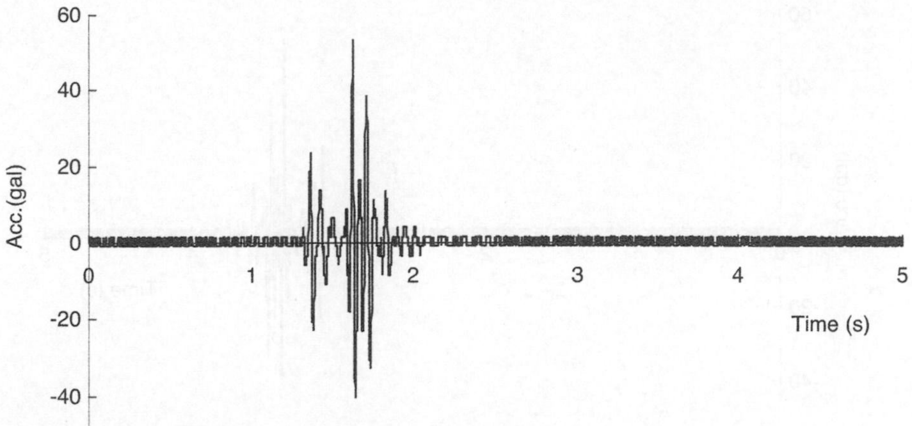

Figure 4. Acceleration of the empty truck (4 ton).

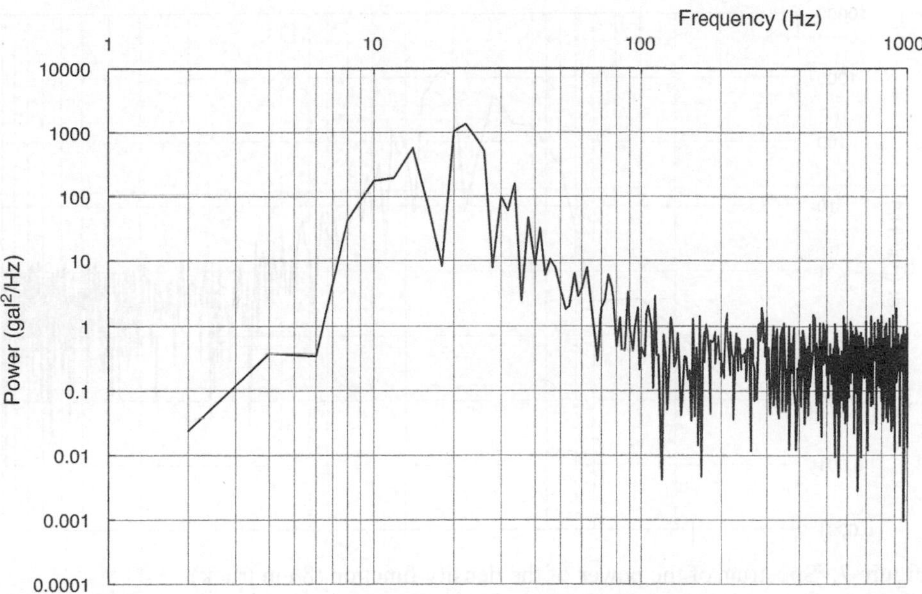

Figure 5. Spectrum of the power of the density function (4 ton truck).

tires and suspension springs of the vehicle. The predominant peaks near 20 Hz at the side of the road are higher than the vibration of the vehicles itself and appear to indicate a different vibration. From 20 Hz to 100 Hz the values of acceleration decrease straightly and the waves at the range higher than 100 Hz maintain constant values larger than the micro tremor. Since the small values measured are disagreeable

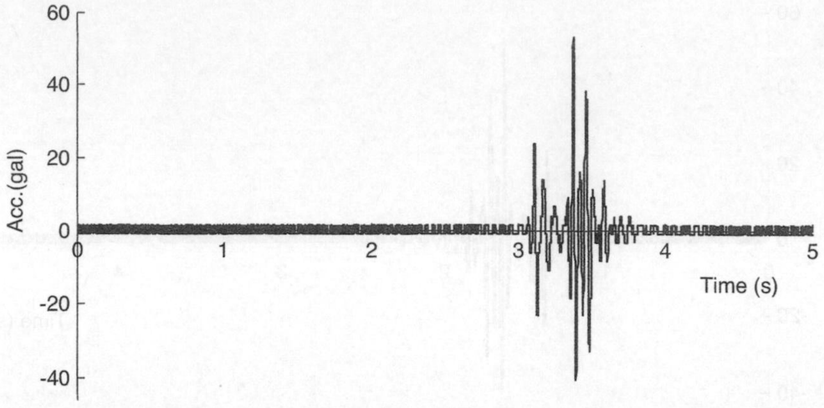

Figure 6. Acceleration of the loaded truck (8 ton).

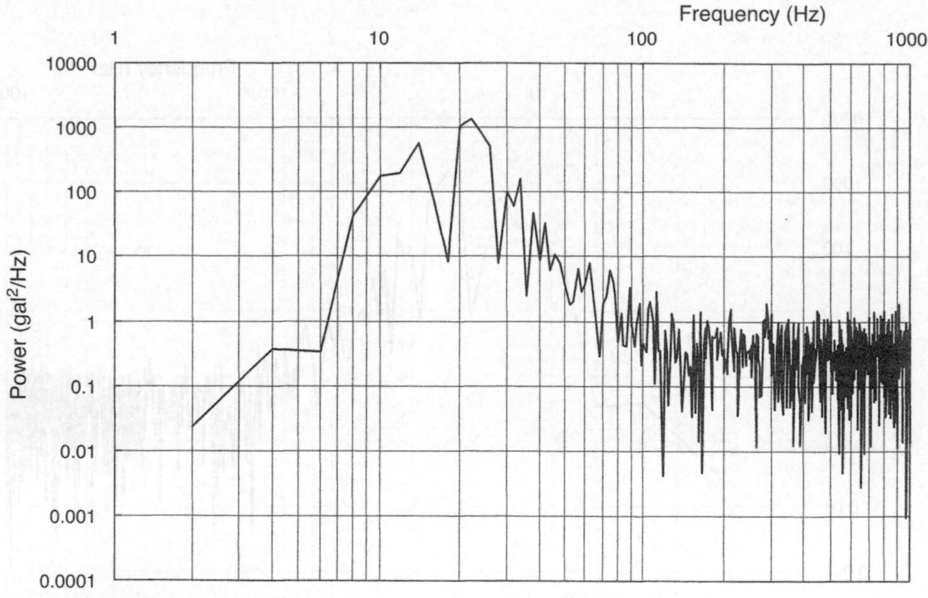

Figure 7. Spectrum of the power of the density function (8 ton truck).

with our actual sensation and it is questionable to promote crushing the pavement, we feel the necessity to perform tests of a wider range of cases using larger holes and steps.

However, it is clear from our past research that the type of force to break the pavement or structures is a sharp shear force. Figure 8 shows the impact wave from a test involving dropping a steel block of 100 N and Figure 9 shows the corresponding wave spectrum. The block fell to a reinforced concrete floor from a height of

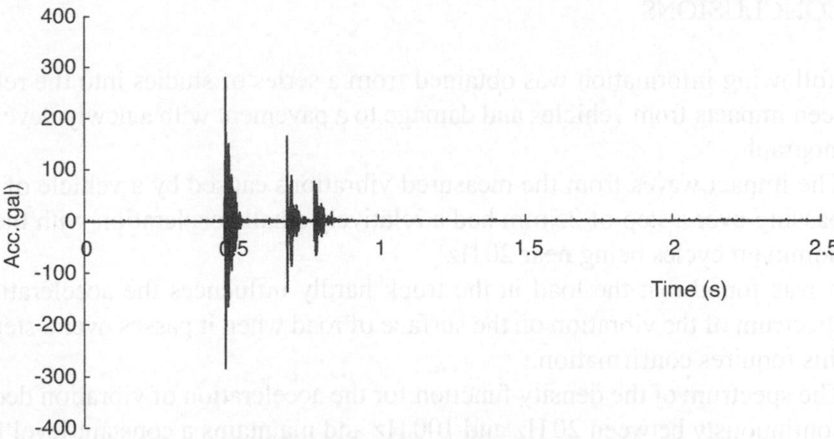

Figure 8. Impact waves from a test involving dropping a 10 kg steel block.

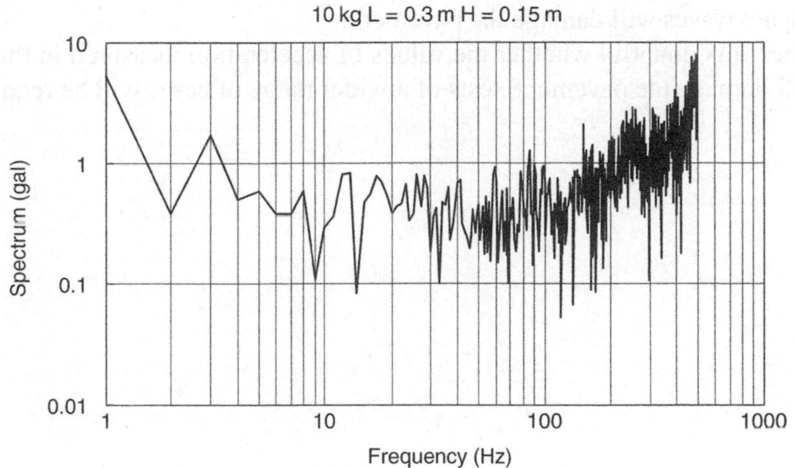

Figure 9. Spectrum of impact waves from a 10 kg steel block.

15 cm 30 cm away from the sensor. The three groups of waves indicate that the block rebounded. The maximum acceleration was about 300 gal and the largest values distributed at the high cycle range of more than 200 Hz. The level of these values was enough to cause damage to pavement if this action repeated. During pile driving, by contrast, the maximum value through the pile is a few hundred G.

Although it was found that the impact from an automobile passing over a small step is not very large, mechanical impacts are large enough that to break and powder surface courses or the aggregate in the base course. The phenomenon of the mud pumping on track bed may be caused by repeated vibration with impact

4 CONCLUSIONS

The following information was obtained from a series of studies into the relation between impacts from vehicles and damage to a pavement with a newly developed seismograph.

(1) The impact waves from the measured vibrations caused by a vehicle of 4 ton passing over a step of 25 mm had a relatively small acceleration with the predominant cycles being near 20 Hz.
(2) It was found that the load in the truck hardly influences the acceleration or spectrum of the vibration on the surface of road when it passes over a step, but this requires confirmation.
(3) The spectrum of the density function for the acceleration of vibration declines continuously between 20 Hz and 100 Hz and maintains a constant level larger than that of micro tremor at cycles higher than 100 Hz.
(4) Although the level of acceleration caused by the vehicle was not very high, given the other examples of impact measured, there is the possibility that repeated impact waves will damage the pavement.
(5) Since it is doubtful whether the values of acceleration measured in this paper will damage the pavement, tests of a wider range of cases will be required.

5

Development and Performance Evaluation of Multi-ring Shear Apparatus

T. Ishikawa & S. Miura
Hokkaido University, Sapporo, Japan

E. Sekine
Railway Technical Research Institute, Tokyo, Japan

ABSTRACT: This paper introduces a new testing method to examine the effects of the rotation of principal stress axes on cyclic plastic deformation of railroad ballast. "Multi-ring shear apparatus" was newly developed as a kind of torsional simple shear test apparatus, and the applicability of the multi-ring shear test to an element test of railroad ballast subjected to moving wheel loads was examined by comparing test results of the multi-ring shear tests with those of model tests. As the results, it was revealed that the multi-ring shear test is appropriate for simulating the actual stress states inside substructures under train loads.

1 INTRODUCTION

The study of "Track deterioration" is one of the principal assignments in railway engineering because track deterioration has serious consequences on the safety of train operation. Track deterioration observed mainly at ballasted tracks as shown in Figure 1 is a phenomenon such that the rail level at train passages is irregular toward the longitudinal direction of railway track with repeated train passages. In general, a dominant factor of track deterioration is supposed to be uneven subsidence of railroad ballast, a pile of well-compacted crushed stones, caused by cyclic wheel loading. The mechanism of track deterioration is being researched in many countries. Nevertheless, there are a number of points uncertain in the mechanism of track deterioration at the present stage. One of the reasons for this seems to lie in the loading methods of conventional model tests and laboratory tests.

So far, a variety of loading tests with model track and model ballast have been performed by many researchers in order to elucidate the cyclic plastic deformation of railroad ballast (Raymond & Bathurst 1987, Kohata et al. 1999). For example, triaxial compression tests with constant amplitude of deviator stress while controlling confining pressure constant have been performed as the element test to elucidate

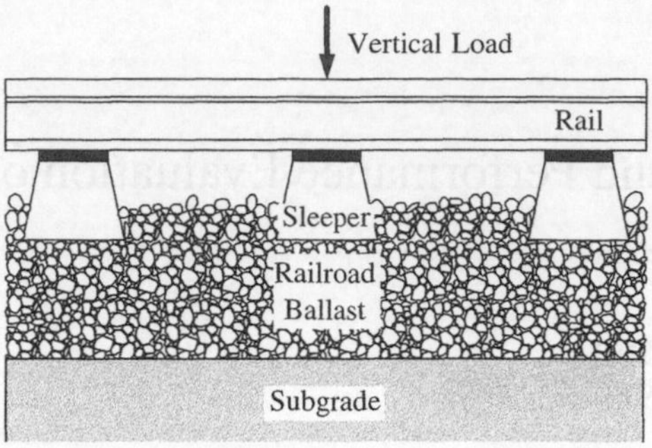

Figure 1. Ballasted track structure.

the behavior of railroad ballast, subjected to moving wheel loads. However, most of these tests are fixed-place cyclic loading tests, in which pulsating compression loads were repeatedly applied to a point of model track or a test sample. In this case, the loads merely increase and decrease, and the loading direction never changes. In contrast, it is said that the principal stress axes rotate inside railroad ballast and subgrade as a train approaches and passes a given location on the rail. Therefore, it is thought that the conventional loading tests cannot simulate the actual stress states inside substructures under train loads, and it seems that the triaxial compression test is not appropriate for simulating the actual stress states inside substructures under train loads because the principal stress axes do not rotate throughout the tests.

This paper introduces a new testing method to examine the effects of the rotation of principal stress axes on cyclic plastic deformation of railroad ballast. First, "Multi-ring shear apparatus" was developed experimentally as a kind of torsional simple shear test apparatus, and the performance was inspected by comparing the test results with those of hollow cylinder torsional shear test. A remarkable feature of the multi-ring shear apparatus is that it can evaluate the effect of rotating principal stress axes under shearing on the strength and deformation characteristics of granular materials. Second, the validity of torsional simple shear tests with the multi-ring shear apparatus for an element test of railroad ballast subjected to moving wheel loads was examined by comparing the test results with those of triaxial compression tests in terms of the effects of the rotation of principal stress axes on cyclic plastic deformation of railroad ballast.

2 TESTING METHODS

2.1 Test implements

Figure 2 shows the schematic diagram of the multi-ring shear apparatus, and it is composed of a bottom plate, a loading plate, and rigid rings that support a specimen.

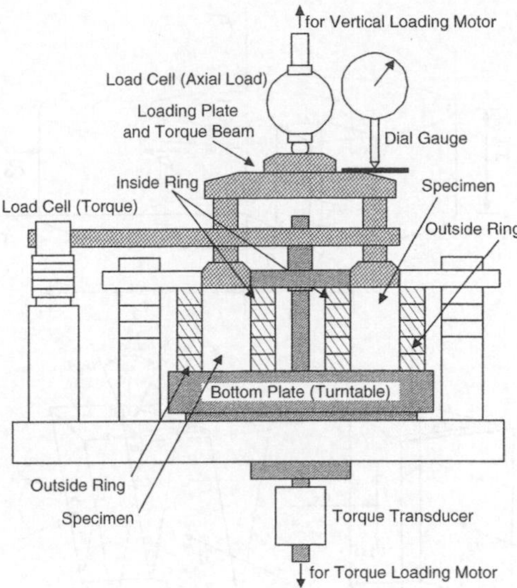

Figure 2. Multi-ring shear apparatus.

The bottom plate supporting a specimen turns by a direct drive motor (DDM) for torque loading though the loading plate is fixed. Consequently, torsion (torque) can be loaded to a specimen confined by the bottom plate, the loading plate, inside rings and outside rings. In addition, vertical loads can be applied to a specimen by the DDM for vertical loading mounted on the loading plate. To decrease the friction between a specimen and rings as much as possible, the structure of inside and outside rings was designed as if each ring can move freely through loading. The width of a specimen was 60 mm (120 mm in inside diameter, 240 mm in outside diameter), and the height is changeable within the range from 40 mm to 100 mm by changing the number of the rings which height is 20 mm. In this paper, the height of a specimen was set equal to 60 mm in order to compare the experimental results obtained from multi-ring shear tests with those of the moving constant loading tests (Ishikawa & Sekine 2002), in which a wheel with constant vertical load travels cyclically along rails, as actual train loading.

The terms regarding size, load, stress and strain used in multi-ring shear tests are defined as shown in Figure 3. The axial stress (σ_a) was measured with a loadcell mounted on the loading plate, and the axial strain (ε_a) was measured with an external displacement transducer (Dial gauge). The shear strain ($\gamma_{a\theta}$) was calculated from the rotation angle of DDM for torque loading and vertical displacement of Dial gauge. In case of monotonic loading tests, the shear stress ($\tau_{a\theta}$) was measured with a torque transducer installed under the bottom plate and a loadcell mounted on the loading plate, though in case of cyclic loading tests, it was mainly measured with the torque transducer. In addition, the lateral pressure (σ_θ) could not be measured due to the problems concerning the measurement precision.

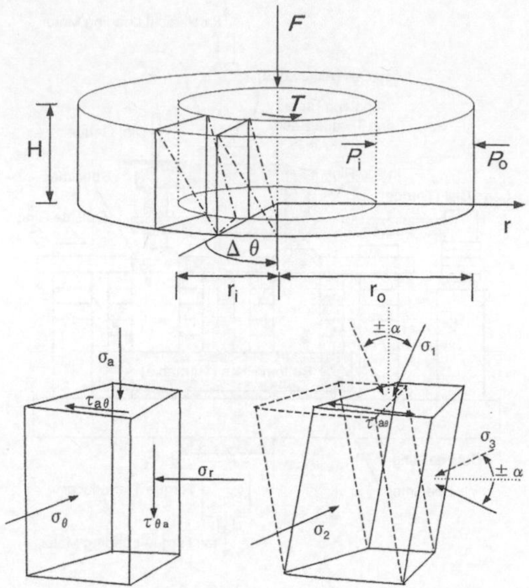

Figure 3. Definition of stress and deformation.

2.2 Test materials and procedures

Two types of test samples which had different mean grain size from each other were employed in this paper. The gradation curves for test samples are shown in Figure 4, together with their mean grain sizes D_{50} and uniformity coefficients U_c. The railroad ballast in Japan is usually composed of angular, crushed, hard andesite stone, namely "ballast." The proper grading of ballast provided by the Japanese railway specification has a grain size distribution between approximately 60 mm and 10 mm. Accordingly, both test samples have one-fifth mean grain size distribution of ballast and the grain size distribution similar to the proper grading of railroad ballast. The term "type A ballast" is used to refer to the model ballast which uniformity coefficient is smaller, and the term "type B ballast" is used to refer to the other. Specimens were prepared by tamping every layer of 20 mm in height with a wooden rammer. The specimens of model ballast were kept under air-dried conditions throughout the tests.

A series of monotonic loading and cyclic loading multi-ring shear tests were individually performed for two types of test samples in order to examine their strength and deformation characteristics. The density of specimens in monotonic loading tests was close to the maximum one, while the density in cyclic loading tests was 90% of the maximum one so as to become close to the density of railroad ballast in the above-mentioned moving constant loading tests. Table 1 summarizes the initial dry densities of all experiment performed in this paper. The loading process was performed as follows. In the monotonic loading tests, after consolidating a specimen of model ballast one-dimensionally under the axial stress (σ_a)

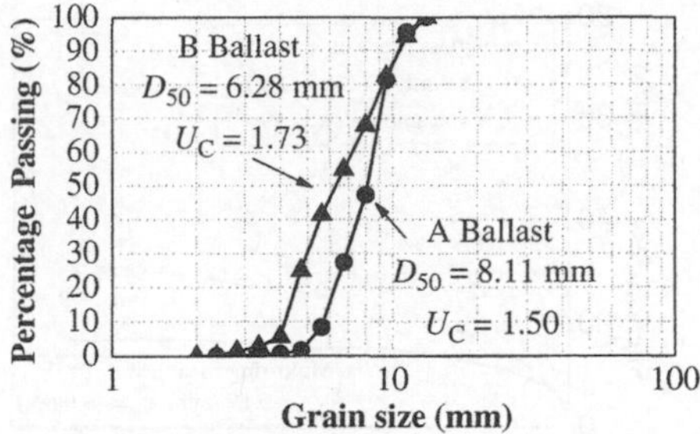

Figure 4. Grain size distribution of ballast.

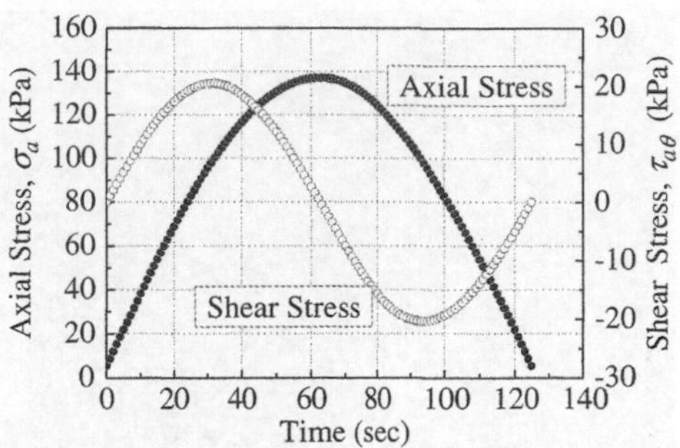

Figure 5. Loading conditions of multi-ring test.

Table 1. Initial dry density of specimen.

Name	Loading condition	Test sample	Density ρ_d
Multi-ring shear test	Monotonic loading	A Ballast	1.54 g/cm^3
Multi-ring shear test	Monotonic loading	B Ballast	1.57 g/cm^3
Multi-ring shear test	Cyclic loading	A Ballast	1.45 g/cm^3
Multi-ring shear test	Cyclic loading	B Ballast	1.51 g/cm^3
Triaxial compression test	Monotonic loading	A Ballast	1.55 g/cm^3
Triaxial compression test	Monotonic loading	B Ballast	1.60 g/cm^3

of 49.0 kPa, the shear stress ($\tau_{a\theta}$) was applied at the constant shear strain rate of 0.01%/min while keeping σ_a constant. In the cyclic loading tests, after one-dimensional consolidation, both $\tau_{a\theta}$ and σ_a in sinusoidal waveforms as shown in Figure 5 were cyclically applied to the specimen. The loading number was 200 cycles, and the loading frequency of 0.008 Hz was selected by referring the experimental conditions of the small scale model tests. Here, the waveform of $\tau_{a\theta}$ and σ_a is the imitation of normal and shear components of the ballast pressure measured in the above-mentioned moving constant loading tests by a two-component loadcell installed between a sleeper and railroad ballast. According to (Raymond & Bathurst 1987), a sinusoidal loading waveform is said to approximate the loading pulse applied to sleepers under actual field conditions.

3 TEST RESULTS AND DISCUSSION

3.1 Performance evaluation of Multi-ring shear apparatus

First, the basic performance of Multi-ring shear apparatus is inspected as a torsional simple shear testing apparatus of granular materials. Figure 6 shows how the specimen of glass beads (D = 5 mm) confined by inside and outside rings

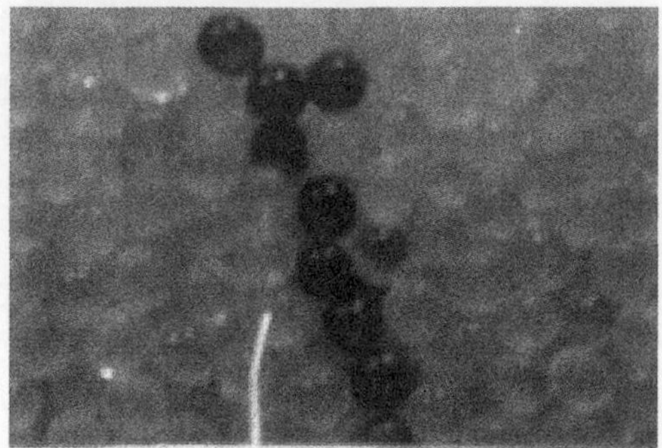

Figure 6. Deformation of glass beads.

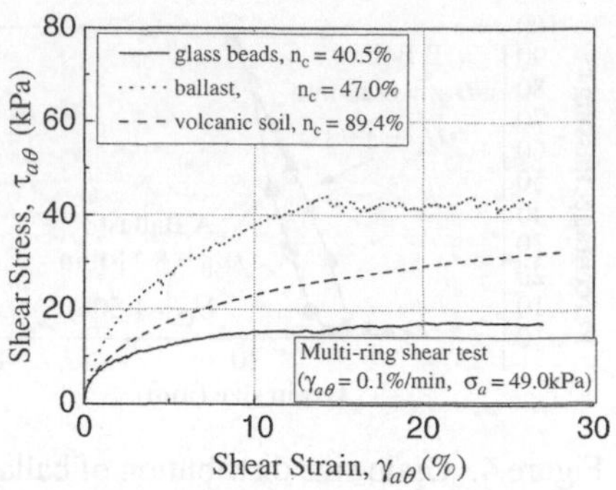

Figure 7. Shear stress–shear strain relations.

deforms under torsional shearing. It can be observed that the horizontal displacement of colored glass beads which line up vertically before shearing linearly increases from the fixed loading plate to the rotated bottom plate. Accordingly, a specimen subjected to torsional shear in multi-ring shear tests is in the simple shear deformation. This indicates that the multi-ring shear apparatus is appropriate for a torsional simple shear testing apparatus of granular materials.

Next, the validity of experimental results obtained from multi-ring shear tests under monotonic loading condition is examined as compared with those of hollow cylinder torisional shear tests performed under the similar experimental conditions. Here, in hollow cylinder torisional shear tests, the size of a specimen was 20 mm in width (60 mm in inside diameter, 100 mm in outside diameter) and 300 mm in height, and the effective confining pressure ($\sigma_c{}'$) was set so as to correspond to the axial stress (σ_a) of 49.0 kPa. Figure 7 shows the relations of various test samples between shear stress ($\tau_{a\theta}$) and shear strain ($\gamma_{a\theta}$) in multi-ring shear tests, and Figure 8 shows the relations in hollow cylinder torisional shear tests likewise. It is recognized that multi-ring shear test results are approximately similar in the shape of stress-stain relationships to hollow cylinder torsional shear test results for both test samples. Moreover, Figure 9 compares the relations of multi-ring shear tests between angle of shear resistance ($\phi_{a\theta}$) and axial stress (σ_a) with the relations of hollow cylinder torsional shear tests between $\phi_{a\theta}$ and $\sigma_c{}'$. In Figure 9, the $\phi_{a\theta}$ increases in order of glass beads, volcanic soils, and ballast regardless of testing methods though $\phi_{a\theta}$ in multi-ring shear tests is smaller than $\phi_{a\theta}$ in hollow cylinder torsional shear tests at the same stress level. These results indicate that the multi-ring shear test is valid as a torsional shear test of granular materials in consideration of the difference in the specimen size, the friction between a specimen and rings and the lateral pressure between both testing methods. Therefore, it seems reasonable to conclude that the mechanical behavior of granular materials under torsional simple shearing can be evaluated with the multi-ring shear apparatus.

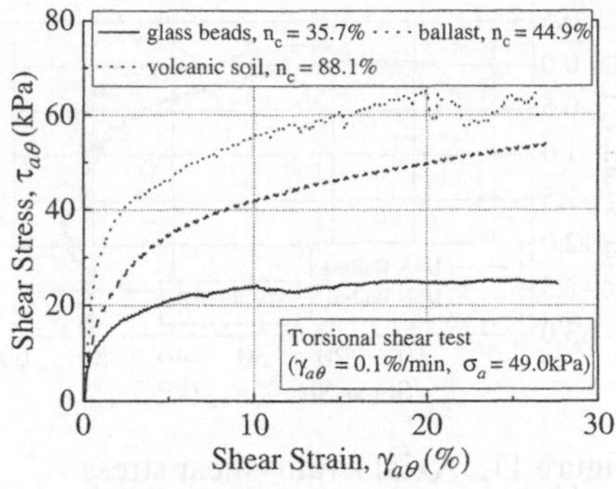

Figure 8. Shear stress–shear strain relations.

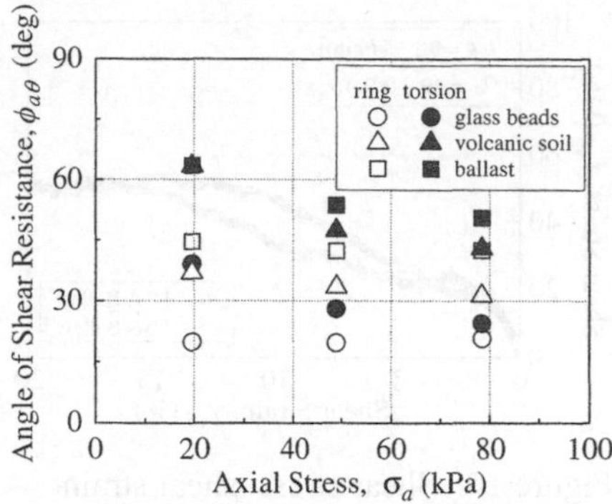

Figure 9. Angle of shear resistance.

3.2 Strength property of ballast

First, the difference in the strength and deformation characteristics between type A and type B ballast is discussed with the multi-ring shear apparatus. Figure 10 shows the relations of both model ballasts between the shear stress $\tau_{a\theta}$ and the shear strain $\gamma_{a\theta}$ in monotonic loading tests. It is observed that the peak strengths of both model ballasts are approximately the same. This indicates that the difference in the grain size distributions has a little influence on the strength property of ballast. Also, Figure 11 shows the relations between the axial strain ε_a and the shear stress $\tau_{a\theta}$. Here, the change of ε_a means the volumetric change of a specimen because both the radial strain (ε_r) and the circumference strain (ε_θ) are zero in the multi-ring shear tests. The volumetric change of model ballast under shearing was altered from compression to dilation at early stages of cyclic loading regardless of the grain size distribution of model ballast. However, the dilation for type A ballast is much larger than type B ballast. This indicates that the slight difference in the coefficient of uniformity has a considerable influence on the deformation behavior of granular materials in the multi-ring shear tests.

Next, discussed is the difference in the strength and deformation characteristics between multi-ring shear tests and triaxial compression tests performed under the similar experimental conditions to multi-ring shear tests. Here, in triaxial compression tests (Kohata & Sekine 2003), the size of a specimen was 150 mm in diameter and 360 mm in height. Figure 12 shows the relations of both model ballasts between the deviator stress and the axial strain in monotonic loading tests, and also Figure 13 shows the relations between the volumetric strain (ε_v) and the deviator stress. Comparing Figure 11 with Figure 13 in terms of the dilatancy under shearing, the effect of the grain size distribution on the dilatancy property of ballast does not appear clearly in triaxial compression tests, while the effect is evident in multi-ring shear tests as described above. Furthermore, in comparing both test results on the maximum angle of shear resistance, the shear resistance derived from triaxial compression test results (Figure 12) is higher than the shear resistance obtained from

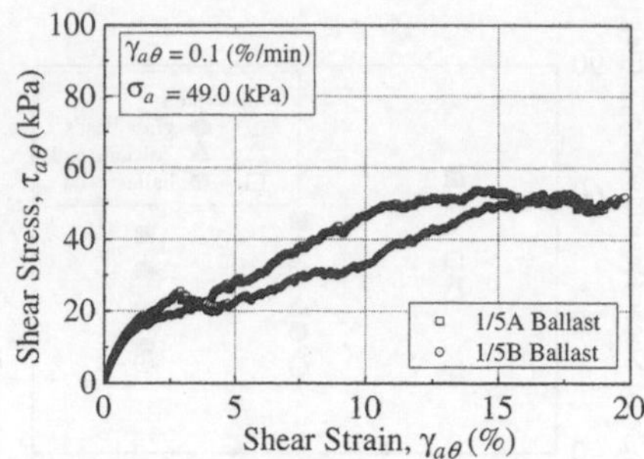

Figure 10. Shear stress–shear strain relations.

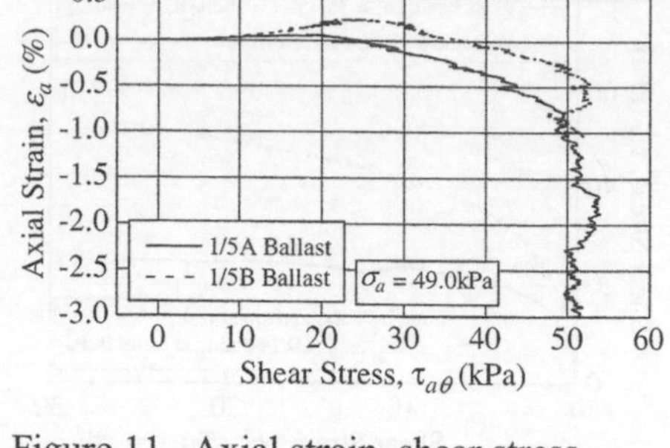

Figure 11. Axial strain–shear stress relations.

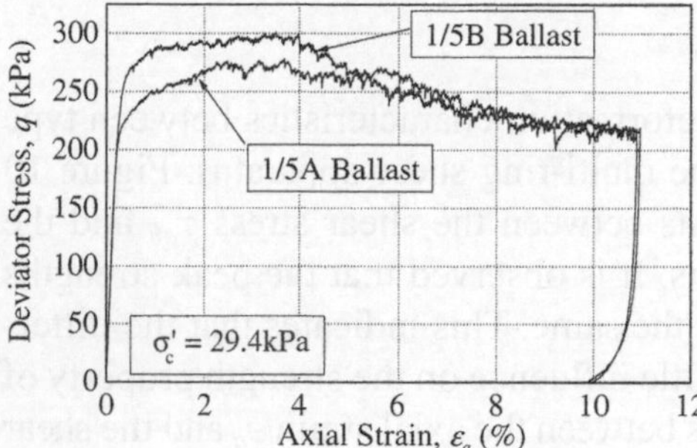

Figure 12. Deviator stress–axial strain relations.

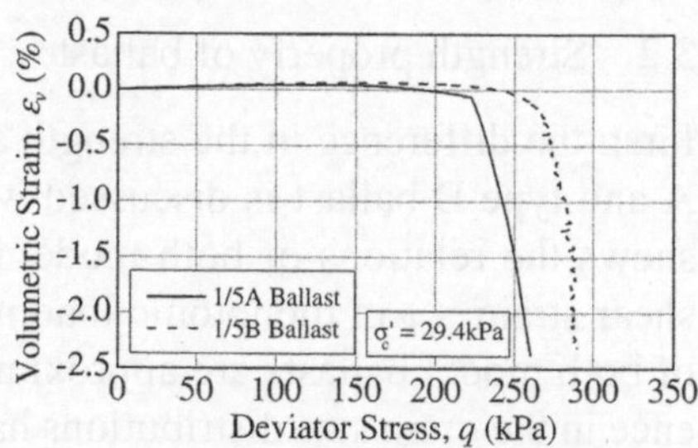

Figure 13. Volumetric strain–deviator stress relations.

multi-ring shear tests (Figure 10) irrespective of the grain size distribution of model ballast. The reason for the difference in the dilatancy property under shearing and the shear resistance between both test results mainly seems to originate in the difference in the specimen size and the boundary condition of a specimen between both testing methods, while the effect of the rotation of principal stress axes observed at multi-ring shear tests can be considered as one of important factors one should not ignore. However, it is difficult from only these experimental results to clarify the effect quantitatively, and further investigation is necessary.

3.3 Cyclic deformation property of ballast

In this section, the applicability of the multi-ring shear tests to an element test of railroad ballast subjected to moving wheel loads was examined. Figure 14 shows the relations of both model ballasts between the average axial strain $(\varepsilon_{ave})_{max}$ at the maximum vertical load, the average axial strain $(\varepsilon_{ave})_{min}$ at the minimum vertical load and the number of loading cycles (N_c) in the above-mentioned moving constant loading tests. Also, Figure 15 shows the relations between the axial strain $(\varepsilon_a)_{max}$ at the maximum axial stress $(\sigma_a)_{max}$, the axial strain $(\varepsilon_a)_{min}$ at the minimum axial

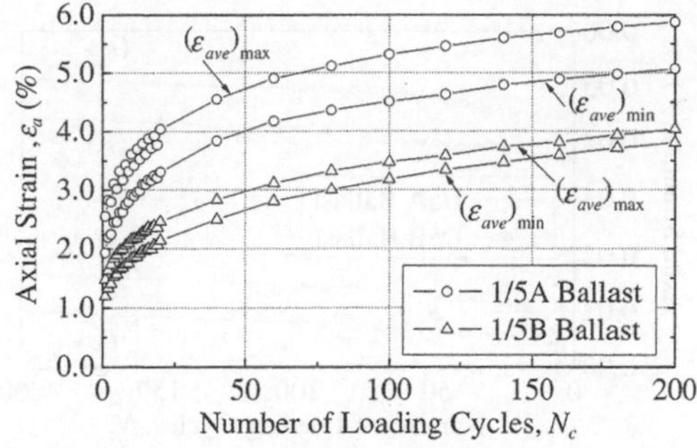

Figure 14. Average axial strain–number of loading cycles relations in moving constant loading tests.

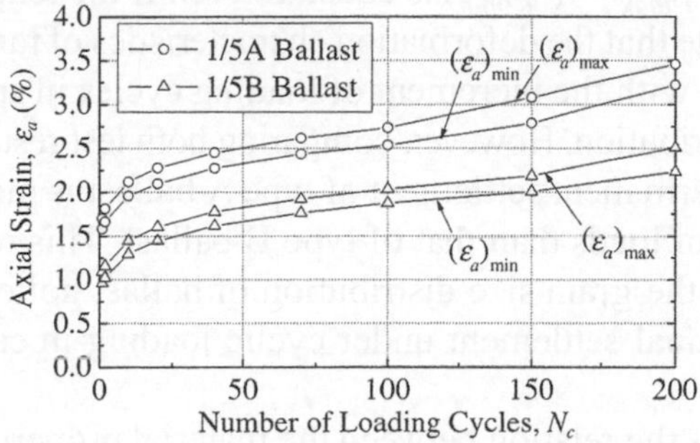

Figure 15. Axial strain–number of loading cycles relations in multi-ring shear tests.

stress $(\sigma_a)_{min}$ and N_c in multi-ring shear tests under cyclic loading condition. Besides, Figure 16 shows the relations between $(\varepsilon_a)_{max}$ at the maximum axial deviator stress, the axial strain $(\varepsilon_a)_{min}$ at the minimum axial deviator stress and N_c in triaxial compression tests under cyclic loading condition. In the moving constant loading tests, the maximum vertical load to a sleeper is generated when a wheel is just above the sleeper, and the minimum vertical load means unloading states as a wheel is far beyond the sleeper. Replacing experimental conditions of the moving constant loading test with those of the multi-ring shear test, $(\sigma_a)_{max}$ may represent the maximum loading state and $(\sigma_a)_{min}$ may represent the unloading state. Therefore, $(\varepsilon_{ave})_{min}$ and $(\varepsilon_a)_{min}$ respectively means the cumulative permanent strain when a wheel load is zero.

First, discussed is the difference in cyclic plastic deformation characteristics due to the difference in grain size distributions of model ballast, on the basis of test results obtained from the multi-ring shear tests. In Figure 15, for both model ballast, both $(\varepsilon_a)_{max}$ and $(\varepsilon_a)_{min}$ increase slowly with the increment of loading cycles N_c after the exponential increment at early stages of cyclic loading, and their rates of increase decrease with the increment of loading cycles. Also, it is recognized that the increase of $(\varepsilon_a)_{max}$ originates in the cumulative permanent deformation of test samples as the

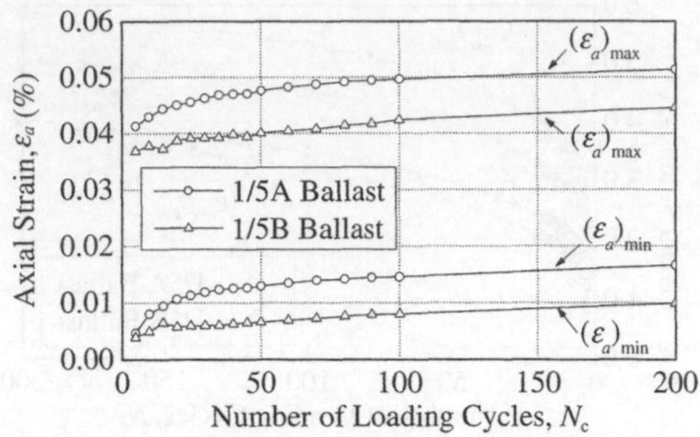

Figure 16. Axial strain–number of loading cycles relations in triaxial compression tests.

strain amplitude $((\varepsilon_a)_{max} - (\varepsilon_a)_{min})$ is constant even if the loading cycle increases. These results indicate that the deformation characteristics of railroad ballast become elastic and constant with the increment of loading cycles in spite of the difference in the grain size distribution. However, comparing both test results from a qualitative point of view, the permanent settlement of type A ballast is more likely to increase with the repetition of loads than that of type B ballast. This result indicates that a slight difference in the grain size distribution of ballast noticeably influences the increase in the residual settlement under cyclic loading in case of the multi-ring shear test.

Next, discussed is the relation between the material properties of ballast and the mechanical properties of railroad ballast by comparing test results of the triaxial compression test or the multi-ring shear test with those of the moving constant loading test. From the comparison of Figure 16 with Figure 14, the accumulated residual strain obtained from cyclic loading triaxial compression tests is very small compared with the macro accumulated residual strain calculated from load-displacement relationships in the moving constant loading tests. Moreover, the difference between test results for two types of model ballast can hardly be distinguished in Figure 16, though the effect of the grain size distribution of ballast on the development of residual settlement of railroad ballast remarkably comes to the surface in employing the moving load instead of the fixed-place cyclic loading as shown in Figure 14. Accordingly, it seems reasonable to conclude that the triaxial compression test with constant amplitude of deviator stress while controlling confining pressure constant is not appropriate for an element test to elucidate the behavior of railroad ballast subjected to train loads.

On the other hands, in comparing Figure 15 with Figure 14, the accumulated residual strain obtained from cyclic loading multi-ring shear tests approximately coincides with the macro accumulated residual strain calculated from test results of the moving constant loading test not only but qualitatively also quantitatively. This demonstrates that the multi-ring shear test has a valuable advantage of excellent applicability to the estimation of deformation behavior of granular materials subjected to repeated moving wheel loads, compared with the triaxial compression test. Therefore, in case of requiring more accuracy of prediction, it is necessary to improve the conventional testing methods for elucidating cyclic plastic deformation of railroad ballast, and for that reason the multi-ring shear test seems to be effective as it can simulate the actual stress states of railroad ballast under train loads.

4 CONCLUSIONS

The following conclusions can be obtained;

- The multi-ring shear apparatus is appropriate for a torsional simple shear testing apparatus of granular materials, and it can evaluate the effect of rotating principal stress axes under shearing on the strength and deformation characteristics of granular materials.
- The strength and dilatancy properties under shearing observed at multi-ring shear tests under monotonic loading condition are different from those at triaxial compression tests, and the effect of the rotation of principal stress axes observed at only multi-ring shear tests in addition to the difference in experimental conditions seems to contribute to the difference.
- A slight difference in the grain size distribution of ballast noticeably influences the increase in the accumulated residual strain under cyclic loading in case of the multi-ring shear test in the same way as real phenomenon, while it has a little influence on the cyclic plastic deformation of ballast in case of the triaxial compression test.
- The newly developed multi-ring shear test has a valuable advantage of excellent applicability to the estimation of deformation behavior of granular materials subjected to repeated moving wheel loads because it can simulate the actual stress states inside sub-structures at train passages, compared with the triaxial compression test with constant amplitude of deviator stress while controlling confining pressure constant.

ACKNOWLEDGMENTS

The authors would like to thank Dr. Yukihiro Kohata, Muroran Institute of Technology for a number of invaluable discussions and suggestions and Mr. Keita Sugiyama, Hokkaido University, who performed laboratory shear tests and arranged the experimental results. This research was supported in part by Grant-in-Aid for Scientific Research (B) from Japan Society for the Promotion of Science (JSPS).

REFERENCES

Ishikawa, T. & Sekine, E. 2002. Effects of Moving Wheel Load on Cyclic Deformation of Railroad Ballast. *Proc. of Railway Engineering-2002, London*,: [1/1(CD-ROM)]

Kohata, Y., Jiang G.L. & Sekine, E. 1999. Deformation characteristics of railroad ballast as observed in cyclic triaxial tests. *Poster Session Proc. of the 11th Asian Regional Conference on Soil Mechanics and Geotechnical Engineering, Seoul, 16–20 August 1999*: 21–22.

Kohata, Y. & Sekine, E. 2003. The effect of cyclic prestraining on the strength and deformation characteristics of the poorly-graded crushed gravels on the similar grain size distribution. *Proc. of Hokkaido Chapter of the Japanese Society of Civil Engineers, 59, Tomakomai, 31 January–1 February 2003*: 514–517 (in Japanese).

Raymond, G.P. & Bathurst, R.J. 1987. Performance of large-scale model single tie-ballast systems, *Transportation Research Record, 1134*: 7–14.

6

Development of Rut Depth Prediction Model Considering Deformation of Asphalt Layer and Subgrade

T. Kanai
Graduate School of Science and Engineering, Department of Civil Engineering, Chuo University, Tokyo, Japan

S. Higashi
Technical Research Institute, Kajima Road Co. Ltd., Tokyo, Japan

K. Matsui
Department of Civil and Environmental Engineering, Tokyo Denki University, Saitama pre., Japan

K. Himeno
Department of Civil Engineering, Chuo University, Tokyo, Japan

ABSTRACT: The life of flexible pavements based on rutting would be estimated using the correlation between number of load applications and permissible subgrade strain. However, few model can be used to predict the rut depth briefly and quickly. Rut depth consists of the permanent deformation in asphalt layer, base course and subgrade. Because the deformation of asphalt layer and subgrade dominate the whole rut depth, they will be employed in this study to predict rut depth. The model based on the creep properties and elastic theory proposed by Usio was applied to predict the deformation of asphalt layer. On the other hand, the deformation of subgrade was estimated by the improved Ushio's equation. To check the accuracy of the proposed prediction model, it was applied to actual rut depth measured by the accelerated load testing facility, LOADSIMULATOR. From findings, the predicted rut depths have good agreement with the measured ones.

1 INTRODUCTIONS

Rutting in a flexible pavement consists of the flow of asphalt layer and the consolidation settlement of base course and/or subgrade. There are many available models (Heukelom 1966, Monismith 1976) to predict rut depth of flexible pavements. With these prediction models, many parameters of material and structure need to be determined. The others need to equip very expensive test machines like a cyclic triaxial test one to obtain the properties of materials. These are subjects to be resolved as soon as possible in order to predict or control rutting of flexible pavements.

In Japan, rutting accounts for rehabilitation of flexible pavements. It is necessary to investigate causes of rutting and predict rut depth. Based on multi-layered elastic theory with the rule of translation between time and temperature, Ushio (Ushio 1982) had developed the sophisticated method to predict the permanent deformation in asphalt layer under repetition of load. He suggested in his model that the rut depth would be underestimated for flexible pavement with granular base, if the permanent deformation were generated in only asphalt layer. In order to predict rut depth correctly, the permanent deformation of base course and subgrade should not be neglected for flexible pavements with granular base.

Because subgrade has the lowest bearing capacity, and tends to contribute the most deformation in a pavement, the predicting model of permanent deformation for subgrade was developed by the transformation of failure criteria equation proposed by shell (Claessen 1977). Then, applying Ushio's predicting method to asphalt layer, rut depth of flexible pavement with granular base can be given by summation of the deformation of both asphalt layer and subgrade. To confirm the effectiveness of proposed predicting model, the comparison was performed for the predicted rut depth with measured one generated by the accelerated load testing facility, LOADSIMULATOR. As a result, the predicted rut depth has good agreement with the measured one. It is found that the proposed model might be useful for rut depth prediction on flexible pavements with granular base course.

2 DEVELOPMENT OF RUT DEPTH PREDICTION MODEL

In this section, the predicting method of subgrade deformation based on Ushio's model is firstly introduced. Also, using the data collected in the site of Japan Highway Public Corporation (JH), it is proved that the accurate rut depth can be predicted for flexible pavements with granular base course by considering the subgarade deformation.

2.1 Ushio's method for prediction of permanent deformation in asphalt layer

The calculation of deformation of asphalt layer consists of following three steps: 1) selecting elastic modulus of each layer considering material properties, temperature and loading time; 2) calculation of the vertical displacements for asphalt layer and subgarade; and 3) evaluation of stiffness of asphalt mixture by creep. Here, the calculation method of permanent deformation in asphalt layer with Equation 1 is explained briefly. The literature (Ushio 1982) offers the detailed information.

$$\Delta H = \frac{S_{mix,D}}{S_{mix,\eta}} \times \delta \tag{1}$$

where, ΔH = permanent deformation in asphalt layer; $S_{mix,D}$ = stiffness of asphalt mixture at short loading time (MPa); δ = compressive deformation of asphalt layer at short loading time (mm); and $S_{mix,\eta}$ = stiffness of asphalt mixture at long loading time (MPa).

STEP 1: The standard load, the short loading time and the standard temperature of asphalt layer should be selected during the analysis. According to these conditions, stiffness of asphalt mixture ($S_{mix,D}$), base and subgrade are evaluated at short loading time.

STEP 2: Using the multi-layered elastic program, GAMES (Maina & Matsui 2002), the compressive deformation of asphalt layer, δ defined the difference of displacements between the top and bottom of asphalt layer can be calculated. Note that the standard load should be divided into several parts considering lateral load wander.

STEP 3: Cumulated (long) loading time is calculated by following three parameters: 1) number of standard load cycles; 2) short loading time; and 3) time-temperature translation factor adjusting to standard temperature of asphalt layer. In Figure 1, read off the stiffness of asphalt mixture at long loading time, $S_{mix,\eta}$ from stiffness of bitumen, S_{bit} obtained by Van der Poel's nomograph (Van der Poel 1954) under cumulated loading time and the standard temperature.

2.2 Development of prediction method of permanent deformation in subgrade

The permanent deformation of base and subgrade besides asphalt layer might be considerable as rutting cause. However, the strain at the top of subgrade is taken as controlling the rut depth in many pavement design methods. The bearing capacity

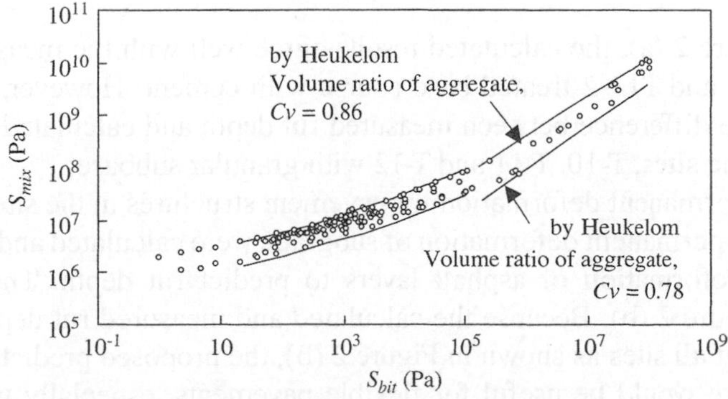

Figure 1. Relationship between S_{bit} and S_{mix} (Ushio 1982).

of subgrade is the lowest in general, the permanent deformation in subgrade would be calculated by the following two procedures in the proposed method.

STEP 4: It is assumed that the cummulated strain is calculated by Equation 2 which is obtained by deformation of shell's equation and the stiffness of subgrade might be equal approximately to vertical stress devided by vertical strain at the top of subgarade.

STEP 5: Next, the permanent deformation at the top of subgrade could be calculated by same method as Equation 1. Equation 3 is obtained from assumption in STEP 4.

$$\varepsilon_{sum} = \varepsilon_{int} \times N^{-0.25} \tag{2}$$

$$\Delta H_{sg} = \frac{S_{sg,D}}{S_{sg,S}} \times d = \frac{\sigma / \varepsilon_{int}}{\sigma / \varepsilon_{sum}} \times d = d \times N^{0.25} \tag{3}$$

where, N = number of standard load cycles (times); ΔH_{sg} = permanent deformation in subgrade (mm); $S_{sg,D}$ = stiffness of subgrade at the short loading time (MPa); $S_{sg,S}$ = stiffness of subgrade at the long loading time (MPa); σ = vertical stress at the top of subgrade (constant in spite of loading time) (MPa); ε_{int} = vertical strain at the top of subgrade at the short loading time; ε_{sum} = cumulated vertical strain at the top of subgrade at the long loading time; and d = compressive deformation at the top of subgrade at the short loading time (mm).

2.3 Example for prediction of rut depth at the site of JH

For the actual tested data shown in Table 1, the permanent deformation of asphalt layer (surface, binder and asphalt treated courses) and subgrade was calculated by proposed model for rut depth prediction. The standard temperature was 22.5°C and the short loading time was 0.02 s. The results are shown in Figure 2 ((a) permanent deformation of only asphalt layers; (b) permanent deformation of asphalt layer and subgrade).

From Figure 2 (a), the calculated results agree well with the measured ones in the sites, T-1 and T13-2 treated base course with cement. However, as rut depth increases, the difference between measured rut depth and calculated one tends to increase in the sites, T-10, T-11 and T-12 with granular subbase.

For total permanent deformation for pavement structures in the sites, T-10, T-11 and T-12, the permanent deformation of subgrade were calculated and added to the permanent deformation of asphalt layers to predict rut depth. The results are shown in Figure 2 (b). Because the calculated and measured rut depths are good agreements at all sites as shown in Figure 2 (b), the proposed prediction model of rut depth here would be useful for flexible pavements, especially with granular base.

Table 1. Data for calculation of rut depth.

	Course	Material and method for construction	T-1	T13-2	T-10	T-11	T-12
Thickness (cm)	Surface	Hot asphalt mixture	4	4	4	4	4
	Binder	Hot asphalt mixture	6	7	14	14	6
	Base	Bituminous stabilization	18	22	–	–	14
		Crushed stone for mechanical stabilization	–	–	20	20	–
		Cement stabilization	17	17	–	–	–
	Subbase	Crusher-run stone	–	–	17	17	23
Stiffness (MPa)	Surface	Hot asphalt mixture	1860	1860	1860	1860	1860
	Binder	Hot asphalt mixture	1860	1860	1860	1860	1860
	Base	Bituminous stabilization	2750	2750	–	–	2750
		Crushed stone for mechanical stabilization	–	–	1540	1400	–
		Cement stabilization	750	1000	–	–	–
	Subbase	Crusher-run stone	–	–	930	440	770
	subgrade	–	46	243	133	208	139

[Remarks]
Poisson coefficient; 0.35 for all layers
Volume ratio of aggregate; Surface and Binder course:0.845, Bituminous stabilized layer: 0.865
Asphalt cement property; Penetration: 62 (1/10 mm), Softening point: 49.5°C

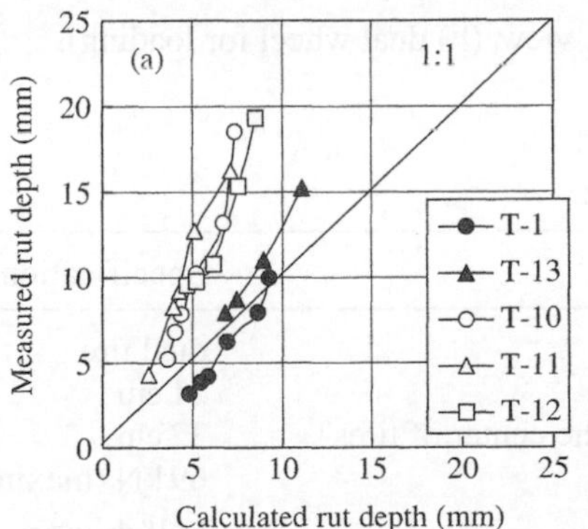

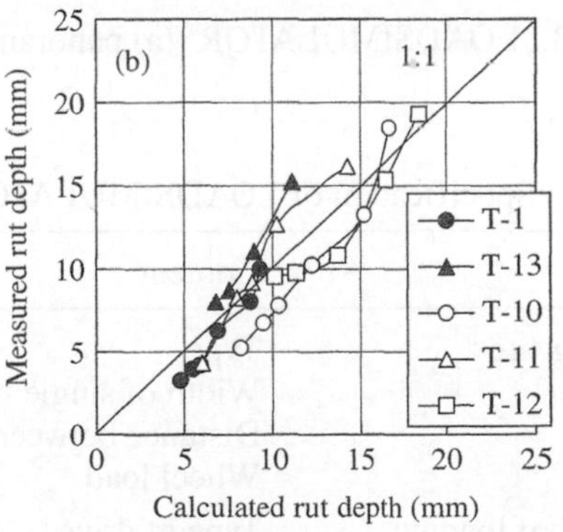

Figure 2. Comparison between predicted rut depth and measured one ((a) only asphalt layers, (b) asphalt layers and subgrade).

3 VALIDATION OF RUT DEPTH PREDICTION MODEL BY LOADSIMULATOR

Authors could confirm the usefulness of proposed rut depth prediction model by literature review. In this section, confirmation is performed using the actual test data by accelerated load testing facility, LOADSIMULATOR.

3.1 Specification of LOADSIMULATOR and condition of accelerated loading test

To investigate performance of pavements during short term, many accelerated load testing facilities (Accelerated pavement testing 1999) are used in the world. In this study, LOADSIMULATOR shown in Figure 3, which was equipped in the machinery center of Kajima Road Co., Ltd. was utilized. The specification of LOADSIMULATOR is shown in Table 2. Note that temperature in the asphalt layer is measured at an hour interval by thermocouple automatically.

Figure 3. LOADSIMULATOR ((a) panoramic view, (b) dual wheel for loading).

Table 2. Specification of LOADSIMULATOR.

Term	Content	Specification
Loading tire	Type	dual tire
	Width of single tire	21 cm
	Distance between the center of tires	32 cm
	Wheel load	69 kN (maximum)
Method of loading	Type of drive	Self driving
	Driving speed	5 km/h
	Width of liateral wander	1 m (maximum)
Test yard	Width	4 m
	Length	30 m

Pavement structure of test yard is shown in Table 3. This asphalt pavement is designed to be able to bear 30,000 passes of 49 kN wheel to fatigue failure by Japanese design method according to "Technical Standard for Pavement Structure". Accelerated loading tests were performed for four seasons from 2004 to 2005. The number of load applications for each season and average temperature of asphalt layer during test are as shown in Table 4.

Dual wheel load was adjusted to be 49 kN. Considering traffic lateral wander, loading test was performed at three patterns shown in Figure 4. The number of load applications by each pattern was uniformly 2500 times for each season.

3.2 Prediction of rut depth

For prediction of rut depth, the standard load was divided into five levels by ratio shown in Figure 5. The vertical displacements at both top and bottom of asphalt layer as well as at the top of subgrade were calculated by GAMES. The loading

Table 3. Pavement cross section at test yard.

Layer	Material	Thickness (mm)	Remarks
Surface	Dense grade asphalt mixture	66	–
Base	Mechanical stabilized aggregate	100	Modified CBR=90%
Subbase	Crusher run stone	100	Modified CBR=90%
Subgrade	Sandy clay	–	Design CBR=8

Table 4. Details of accelerated loading test.

Season	Test date	Number of load application (times)	Average temperature in asphalt layer (°C)
Spring	June, 2004	7500	31.7
Summer	August, 2004	7500	34.8
Fall	November, 2004	7500	12.8
Winter	February, 2005	7500	5.7
Total		30,000	–

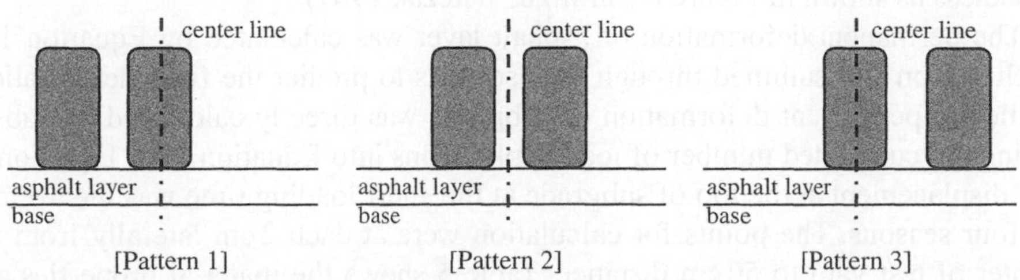

Figure 4. Lateral wander pattern of wheel pass (three patterns).

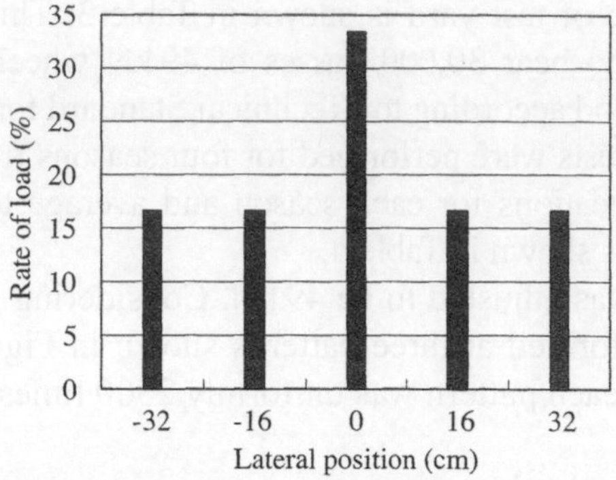

Figure 5. Lateral distribution of load.

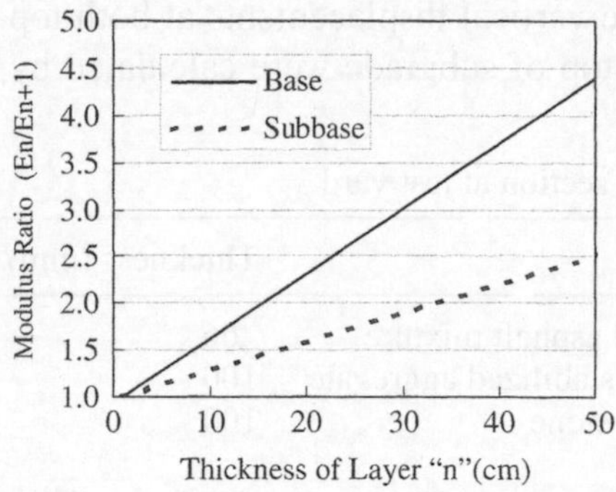

Figure 6. Relationship between thickness and modulus ratio (base and subbase).

contact area was circular with a radius of 12.9 cm to be same as the measured contact area, 525 cm^2.

Stiffness of asphalt mixture, $S_{mix,D}$ at the short loading time and $S_{mix,\eta}$ at the long (cumulated) loading time were read off from the relationship between S_{bit} and S_{mix} shown in Figure 1. The upper limit line in Figure 1 was applied to determine S_{mix} from S_{bit}, because volume ratio of aggregate, Cv was approximately 0.86. Stiffness of subgrade was 80 MPa (i.e. 10×design CBR). Stiffness of base and subbase were calculated by modulus ratio to underlayer decided from the layer thickness as shown in Figure 6 (Smith & Witczak 1981).

The permanent deformation of asphalt layer was calculated by Equation 1 at each season and summed through four seasons to predict the final deformation, while the permanent deformation of subgrade was directly calculated by substituting the cumulated number of load applications into Equation 3. In Equation 3, the displacement at the top of subgrade at the short loading time was the average of four seasons. The points for calculation were at each 2 cm laterally from the center of test yard to 50 cm distance. Table 5 shows the material properties and loading condition at accelerated loading test.

Table 5. Material properties and loading condition at accelerated loading test.

Term		Unit	Spring	Summer	Fall	Winter
Test date		–	June, 2004	August, 2004	November, 2004	February, 2005
Stiffness at short time loading	Asphalt mixture	MPa	440	303	4278	10058
	Base course	MPa	171	171	171	171
	Subbase course	MPa	102	102	102	102
	Subgrade	MPa	80	80	80	80
Average of temperature in asphalt layer		°C		31.7	34.8	12.8 5.7
Number of load applications		time	7500	7500	7500	7500
Loading time		s	1350	1350	1350	1350
Stiffness at long time loading	Asphalt cement	Pa	1.516×10^2	7.617×10	2.584×10^4	1.700×10^5
	Asphalt mixture	MPa	9.0	7.3	37.1	152.3
Stiffness ratio of asphalt mixture (short/long)		–	49	41	115	66

[Remarks]
Poisson coefficient: 0.35 for all layers except for subgrade (0.4)
Aggregate volume: 0.86, Loading time per pass: 0.18 (s)
Asphalt cement property; Penetration: 46 (1/10 mm), Softening point: 50.0°C (assumed from empirical experience)

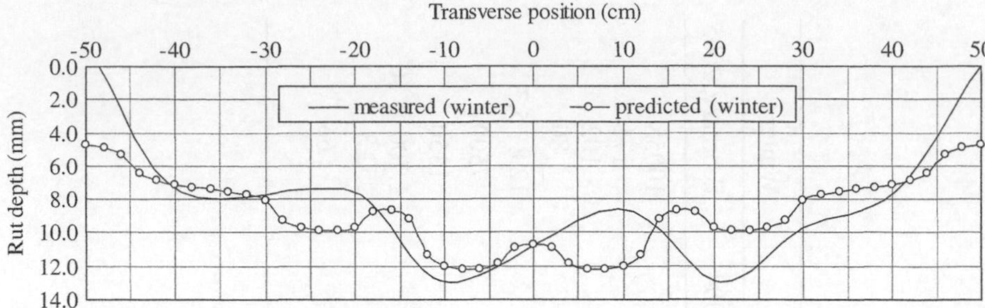

Figure 7. Comparison of predicted profile and measured one in the winter.

3.3 Results of rut depth prediction

Transverse profiles were measured at four lines. For example, the calculated profile is compared with the measured one at No.1-line in the winter in Figure 7. From Figure 7, the position where the maximum of deformation generates at the right side is different between the calculated and the measured, however, the maximum of permanent deformation is almost same each other at both sides. There was no cracking in the test yard at the end of accelerated loading test.

On the other hand, the measured rut depths (maximum of permanent deformation) progress as shown in Figure 8(a). Figure 8(b) shows the comparison between the average of measured rut depth at four lines and the predicted rut depth. From Figure 8(b), the deformation of asphalt layer increases in the spring (7500 times) and summer (15000 times) when temperature is relatively high, and the deformation of subgrade remarkably increases until number of load applications, 7500 times, thereafter it progresses slowly at constant rate. Finally, it is found that because the predicted rut depth is almost same as the measured one, the prediction result by suggested model might be reasonable.

4 CONCLUSIONS

Conclusions in this study are as following;

1) We developed the rut depth prediction model on asphalt pavement with granular base, by combining the prediction method for permanent deformation of subgrage to Ushio's prediction method for asphalt layer.
2) By suggested rut depth prediction model, it could be simulated that the permanent deformation of asphalt layer increases in the spring and summer when temperature is relatively high, and the deformation of subgrade remarkably increases at the beginning of test, on accelerated loading test of asphalt pavement with granular base.

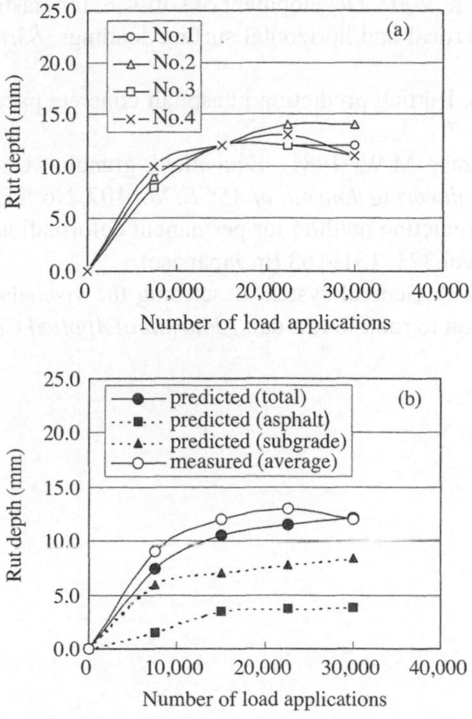

Figure 8. Comparison of predicted rut depths and measured ones ((a) measured rut depth, (b) measured vs. predicted).

3) Also, the validation of this rut depth predicting model for asphalt pavement with granular base might be confirmed, because the predicted rut depth almost agrees with the measured one on the results of accelerated loading test.
4) It should be necessary to perform accelerated loading test at other type of asphalt pavement, especially with thicker asphalt layer for more verification of this prediction model.

REFERENCES

Accelerated Pavement Testing 1999. *International Conference of Accelerated Pavement Testing*. Nevada.

Claessen, A.I.M. Edwards, J.M. & Uge, P. 1977. Asphalt pavement design – shell method, *4th International Conference on the Structural Design of Asphalt Pavements, Michigan*: 39–74.

Heukelom, W. 1966. Observation on the rheology and fracture of bitumens and asphalt mixes, *Proceeding of AAPT*, Vol.35.

Maina, J.W. & Matsui, K. 2002. Development of software for elastic analysis of pavement structure due to vertical and horizontal surface loadings, *83rd TRB Annual meeting, Washington D.C.*

Monismith, C.L. 1976. Rutting prediction in asphalt concrete pavements, *TRB, Research record*, Vol.616: 2–7.

Smith, B.E. & Witczak, M.W. 1981. Equivalent granular base moduli: prediction, *Transportation Engineering Journal of ASCE*, Vol.107, No. TE6: 635–652.

Ushio, S. 1982. The predicting method for permanent deformation of asphalt pavements, *Journal of JSCE*, Vol.323: 151–163 (in Japanese).

Van der Poel, C. 1954. A General system describing the viscoelastic properties of bitumens and its relation to routine test data, *Journal of Applied Chemistry*.

7

Effects of Continuous Principal Stress Axis Rotation on the Deformation Characteristics of Sand Under Traffic Loads

Y. Momoya, K. Watanabe, E. Sekine & M. Tateyama
Railway Technical Research Institute, Tokyo, Japan

M. Shinoda
Integrated Geotechnology Institute Limited, Tokyo, Japan

F. Tatsuoka
Tokyo University of Science, Chiba, Japan

ABSTRACT: Large continuous rotation of the principal stress axis has a significant effect on the residual deformation of ground subjected to cyclic loading, such roadbed and subgrade under traffic load. To investigate its effect, a new special stress-strain test apparatus that can control at will the principal stress axis direction under plane strain conditions was developed based on the conventional simple shear test system. The following four types of cyclic loading test with different patterns of principal stress axis rotation were carried out; a) without principal stress axis rotation; and with principal stress axis rotation; b) simulating the one in railway track; c) simulating the one in the ground beneath a compaction roller; and d) keeping the stress state always traveling along a fixed Mohr's circle of stress. The test results showed that the residual deformation by cyclic loading of sand and gravelly soil increases with the magnitude of principal stress axis rotation. The effects in test b) were not negligible, but much smaller than those in tests c) and d) under otherwise the same test conditions.

1 INTRODUCTION

To investigate the instantaneous and residual deformation of railway roadbed under trainload, fixed-point cyclic loading tests, applying repeated load to the same point on rails, are often carried out, because it is usually very difficult to perform full-scale moving-wheel loading tests. For example, the deformation of solid bed track constructed on an asphalt pavement (Momoya et al., 2002) and the residual settlement of a sleeper on a slag-reinforced roadbed (Hwang et al., 2001) were investigated by performing full-scale fixed-point cyclic loading tests. However, the stress conditions in railway roadbed and subgrade under fixed-point loading conditions are significantly different from those when subjected to moving-wheel load under train running, in which constant wheel load moves continuously on rails (Fig. 1). To investigate the effect of these different loading conditions, Hirakawa et al. (2002) carried out both moving-wheel and fixed-point cyclic loading tests on a scale model of railway track arranged on a subgrade of air-dried Toyoura sand. Momoya & Sekine (2005) carried out similar but more sophisticated tests on a railway asphalt roadbed arranged on a subgrade of gravely sand. Those test results showed that the residual settlement of sleeper by repeated moving-wheel loading is much larger than the one by fixed-point cyclic loading.

There are two major factors for the difference in the settlement of sleeper between the two different loading conditions. The first one is different deformation modes of railway track system under the different loading conditions. Under fixed-point cyclic loading, the instantaneous settlement and therefore the residual settlement of the sleeper at the fixed cyclic loading point inevitably become larger than those of the adjacent sleepers. As a result, due to a rigidity of rail, even when the load amplitude and the maximum and minimum loads on rails are kept constant, the maximum load applied to the sleeper above which fixed-point cyclic load is applied gradually decreases whereas the maximum load applied to the adjacent sleepers gradually increases. Therefore, the residual settlement of the sleeper beneath the instantaneous loading point becomes eventually much smaller under fixed-point loading than under moving wheel loading. Momoya & Sekine (2004) confirmed this mechanism by performing scale model tests under both loading conditions.

The second factor is the continuous rotation of principal stress axis under moving-wheel loading, which is not the case under fixed-point cyclic loading. Wong & Arthur (1984) investigated the effects of this factor on the residual deformation of sand by continuously rotating the directions of σ_1 and σ_3 keeping the sin ϕ_{mob} value constant

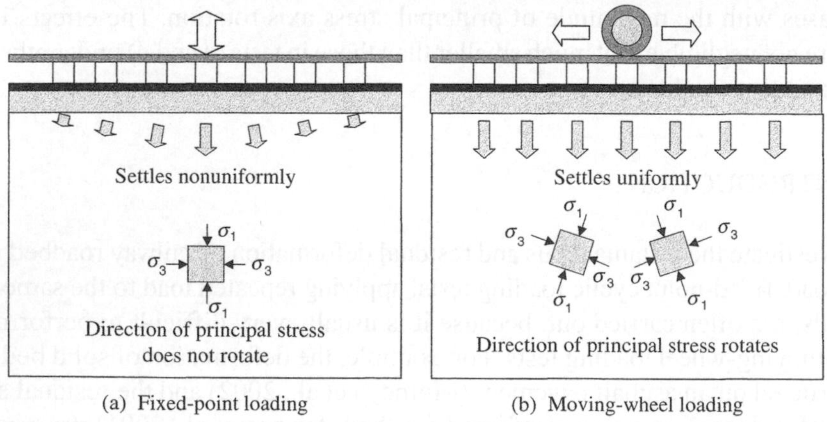

Figure 1. Schematic view of different stress conditions in subgrade between fixed-point loading and moving-wheel loading.

in a cubic specimen arranged in the directional shear cell. Towhata et al. (1994) carried out torsion shear tests on a hollow cylindrical specimen of sand to simulate stress paths in the subgrade under train load assumed to be strip load. These studies revealed that the effects of this factor become larger with an increase in the magnitude of principal stress axis rotation. It seems however that the direct shear cell is extremely complicated to produce and operate. In hollow cylinder tests, it is very difficult to control the radial thickness of specimen, as the lateral boundary of specimen is stress-controlled. For this reason, it is very difficult to keep the plane strain conditions, which is the most basic strain condition that should be referred to when investigating the present issue. Furthermore, unless the specimen becomes very large (say 100 cm in the outer diameter), a hollow cylindrical specimen cannot accommodate large grain size materials, such as crushed stone as used in the railway roadbed and subgrade.

In view of the above, the authors developed a new stress-strain test apparatus that can control the principal stress axis direction under plane strain conditions while it can accommodate gravel soils (Fig. 2). The new apparatus was designed based on the conventional simple shear test system aiming at a high operationality,

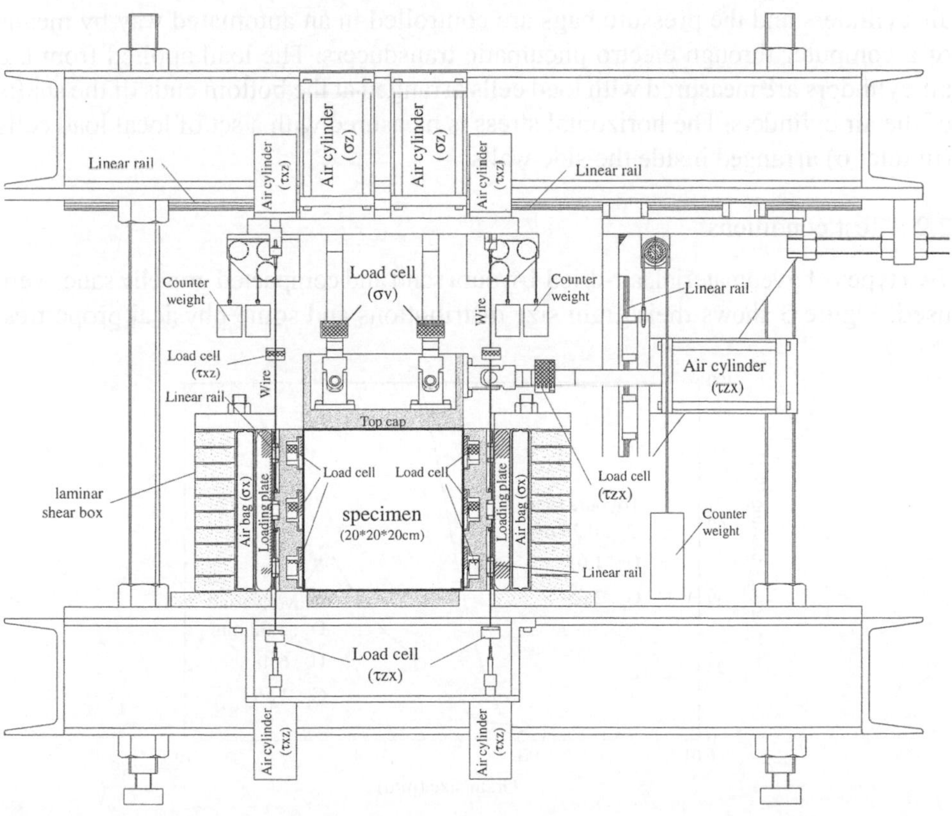

Figure 2. A new apparatus that can control the principal stress axis rotation.

as explained in the next section. To evaluate the performance of the new apparatus, a series of cyclic loading tests with different patterns of principal stress axis rotation were performed using a fine uniform sand (Toyoura sand) and a gravelly sand.

2 TEST METHOD

2.1 A new apparatus controlling principal stress axis rotation

The specimen is cubic (20 cm × 20 cm × 20 cm), which can accommodate materials for subgrade and roadbed, including gravels with 20–30 mm grain size. The specimen is set in the laminar shear box, divided in 10 sub-layers. The vertical normal and horizontal shear stresses, σ_z and τ_{zx}, on the top cap and the vertical shear stress, τ_{xz}, on the side walls are independently applied by using a set of air cylinders (i.e., Bellofram® cylinders); i.e., four to control σ_z, two to τ_{zx} and to control τ_{xz}. The horizontal normal stress, σ_x, on the side walls is applied by means of two air pressure bags, set immediately behind the side walls. The air pressures in the air cylinders and the pressure bags are controlled in an automated way by means of a computer through electro pneumatic transducers. The load applied from the air cylinders are measured with load cells arranged at the bottom ends of the shafts of the air cylinders. The horizontal stress is measured with a set of local load cells (in total 6) arranged inside the side walls.

2.2 Test conditions

Two types of gveomaterial, air-dried Toyoura sand and compacted gravelly sand, were used. Figure 3 shows their grain size distributions and some physical properties.

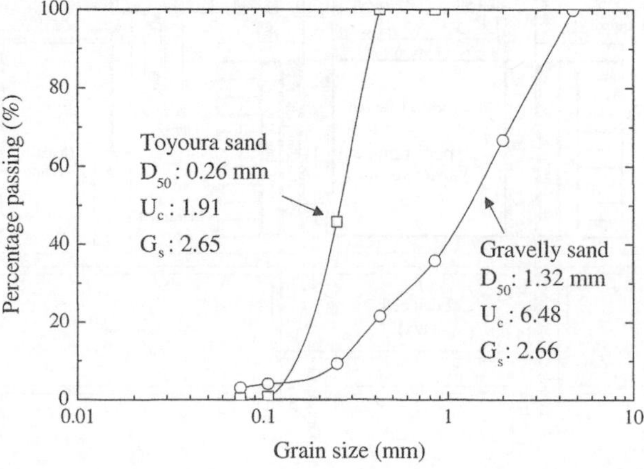

Figure 3. Grain size distributions and physical properties of Toyoura sand and gravelly sand.

The dry density of Toyoura sand specimen was 1.56 g/cm^3 (D_r = 80%) and that of gravelly sand specimen was 1.66 g/cm^3 (D_c = 85%).

The following four types of cyclic loading test with different patterns of principal stress axis rotation, as described in figure 4, were carried out:

Test (a): Without principal stress axis rotation with the directions of principal stress axis fixed to be vertical during cyclic loading.

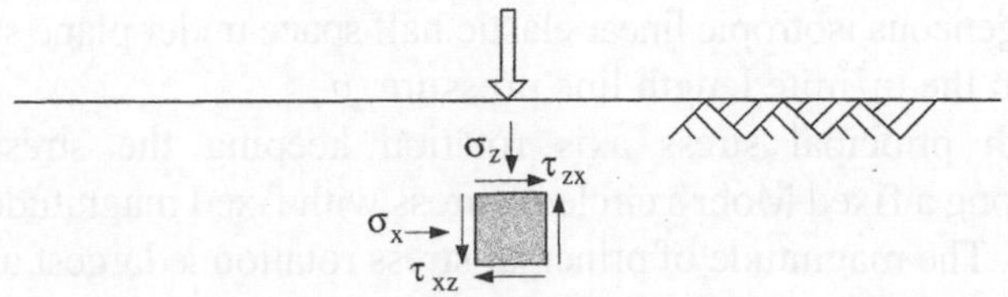

(a) Without principal stress axis rotation

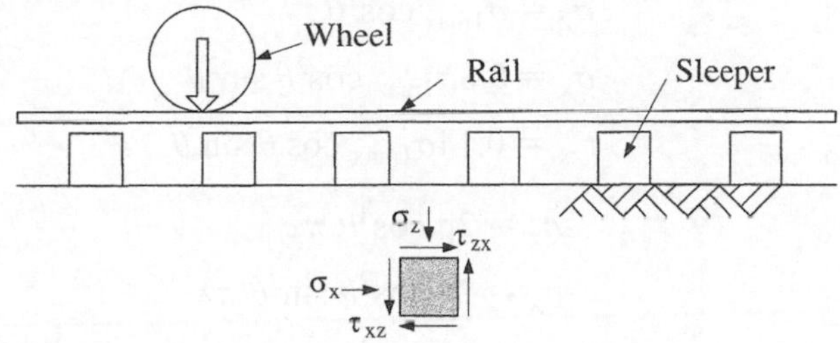

(b) Principal stress axis rotation under railway track

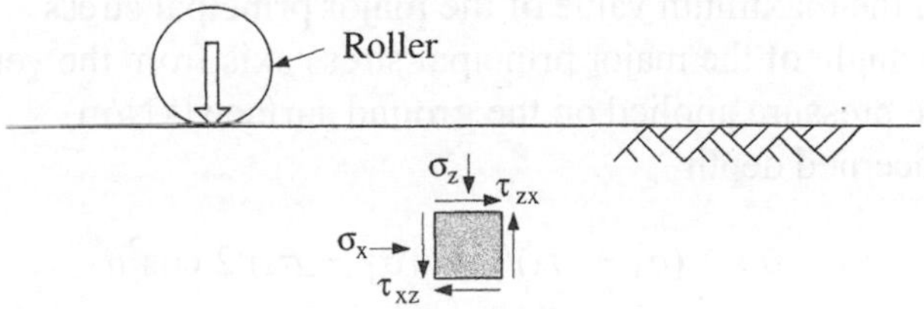

(c) Principal stress axis rotation under a compaction roller

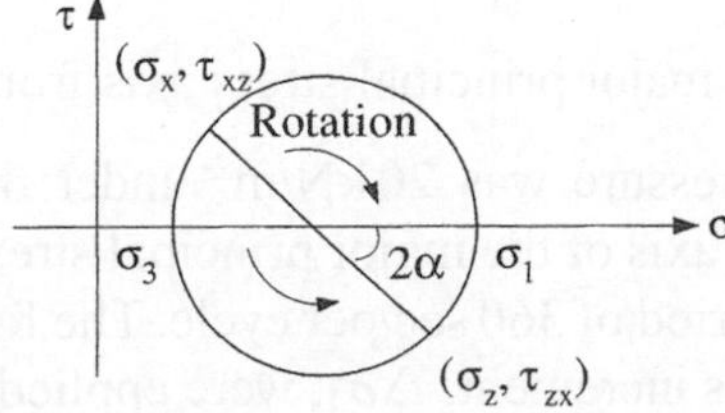

(d) Principal stress axis rotation along a fixed Mohrís circle of stress

Figure 4. Stress conditions in cyclic loading tests with and without principal stress axis rotation.

Test (b): With principal stress axis rotation simulating the one in railway track with relatively small principal stress rotation. The stress history applied to the specimen was determined by an empirical equation Eq. (1a)–(1c), which was obtained from the results of the moving-wheel loading test carried out by Momoya & Sekine (2005).

Test (c): With principal stress axis rotation simulating the one in the ground beneath a compaction roller with relatively large principal stress axis rotation. The applied stress history is given by Eq. (2a)–(2c), the Bousinesseq solution for a homogeneous isotropic linear elastic half space under plane strain conditions subjected to the infinite length line pressure, p.

Test (d): With principal stress axis rotation keeping the stress state always traveling along a fixed Mohr's circle of stress with fixed magnitudes of the principal stresses. The magnitude of principal stress rotation is largest among the cases examined in the present study. The applied stress history is given by Eq. (3a)–(3c).

$$\sigma_z = \sigma_{1max} \cos^2\theta \tag{1a}$$

$$\sigma_x = 0.8\sigma_{1max} \cos^2\theta \sin^2\theta \tag{1b}$$

$$\tau_{zx} = 0.14\sigma_{1max} \cos\theta \sin\theta \tag{1c}$$

$$\sigma_z = 2p \cos^4\theta/\pi z \tag{2a}$$

$$\sigma_x = 2p \cos^2\theta \sin^2\theta/\pi z \tag{2b}$$

$$\tau_{zx} = 2p \cos^3\theta \sin\theta/\pi z \tag{2c}$$

where σ_{1max}: the maximum value of the major principal stress
θ: the angle of the major principal stress axis from the vertical axis
p: line pressure applied on the ground surface (kN/m)
z: concerned depth

$$\sigma_z = (\sigma_1 + \sigma_3)/2 + (\sigma_1 - \sigma_3)/2 \cos^2\theta \tag{3a}$$

$$\sigma_x = (\sigma_1 + \sigma_3)/2 - (\sigma_1 - \sigma_3)/2 \cos^2\theta \tag{3b}$$

$$\tau_{zx} = (\sigma_1 - \sigma_3)/2 \sin^2\theta \tag{3c}$$

where θ: the angle of the major principal stress axis from the vertical axis

The initial confining pressure was 20 kN/m^2 under the isotropic conditions. In tests (b), (c) and (d), the axis of the major principal stress, σ_1, was continuously rotated cyclically with a period of 360 sec per cycle. The following two amplitudes of the major principal stress increment, $\Delta\sigma_1$, were applied in tests (a) through (d):

$$\text{Case A: } \Delta\sigma_1 = 10.61\,\text{kN/m}^2 \tag{4a}$$

$$\text{Case B: } \Delta\sigma_1 = 21.22\,\text{kN/m}^2 \tag{4b}$$

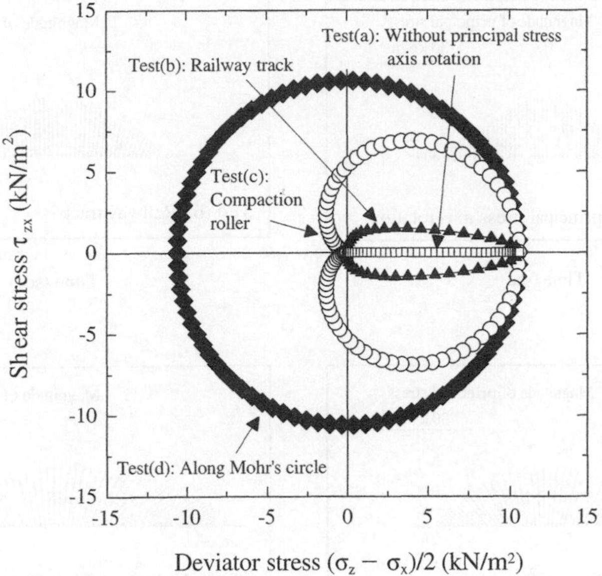

Figure 5. Stress paths on the deviator stress-shear stress plane in four types of test (Case B, $\Delta\sigma_1 = 21.22\,\text{kN/m}^2$).

Cases A & B in test (c) correspond to the stresses at a depth $z = 0.3\,\text{m}$ under line pressure, respectively, $p = 5\,\text{kN/m}$ and $p = 10\,\text{kN/m}$. Figure 5 shows the stress paths on the deviator stress – shear stress plane and figure 6 shows the direction of the σ_1 axis and its magnitude in the four types of test in Case B ($\Delta\sigma_1 = 21.22\,\text{kN/m}^2$).

3 TEST RESULTS

3.1 Stress histories applied to the specimen

Figure 7 is a typical comparison between the input data and measured stresses from a test (c) with $\Delta\sigma_1 = 21.22\,\text{kN/m}^2$. The following trends of behaviour may be seen:

1) The vertical and shear stresses, σ_z and τ_{zx}, on the cap, measured with load cells agreed well with the input data.
2) However, the horizontal stress, σ_x, measured with a set of local load cells arranged inside the side walls drifted gradually at a low rate with cyclic loading. This stress drift may be due to the development of the non-uniform stress distribution in the specimen with cyclic loading.
3) The increment of the major principal stress, $\Delta\sigma_1$, was kept nearly constant as intended.

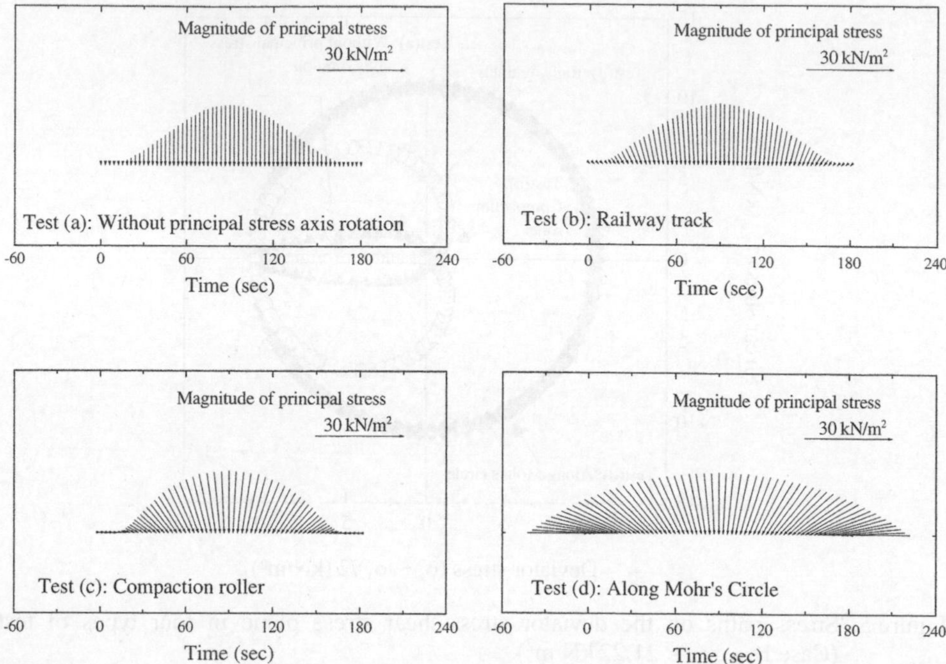

Figure 6. Directions and magnitude of the major principal stress in the four tests in Case B ($\Delta\sigma_1 = 21.22\,\text{kN/m}^2$ (Note: Considering the own weight of soil, the direction of principal stress is not exactly -90 deg $-$ 0deg $-$ $+90$ deg).

Because of the factor above, however, $\Delta\sigma_1$ also exhibited a gradual drift from the input data, although it was at a low rate.

To alleviate this stress drift problems, it is necessary to introduce a feedback control system for the horizontal stress. Despite this defect, it may be seen that the stress condition in the specimen was generally very well controlled.

3.2 Residual strains by cyclic loading with and without principal stress axis rotation

Figure 8 shows the residual vertical strain, which is equal to the residual volumetric strain, $(\varepsilon_v)_{res}$, developed in the cyclic loading tests with and without principal stress axis rotation. Figures 8a & b are for Toyoura sand with an amplitude of the major principal stress $\Delta\sigma_1 = 10.61\,\text{kN/m}^2$ and $21.22\,\text{kN/m}^2$, while Figs. 8c for gravelly sand with $\Delta\sigma_1 = 21.22\,\text{kN/m}^2$. With Toyoura sand, $(\varepsilon_v)_{res}$ was largest in test (d) (along a fixed Mohr's circle of stress), while it was second largest in test (c) (compaction roller). With gravelly sand, on the other hand, $(\varepsilon_v)_{res}$ was largest in test (c) and second largest in test (d). The reason for this discrepancy is not known to the present authors. It is to be noted that, in test (d) (along a fixed Mohr's circle of stress), the residual strain by cyclic loading is considerable due to continuous

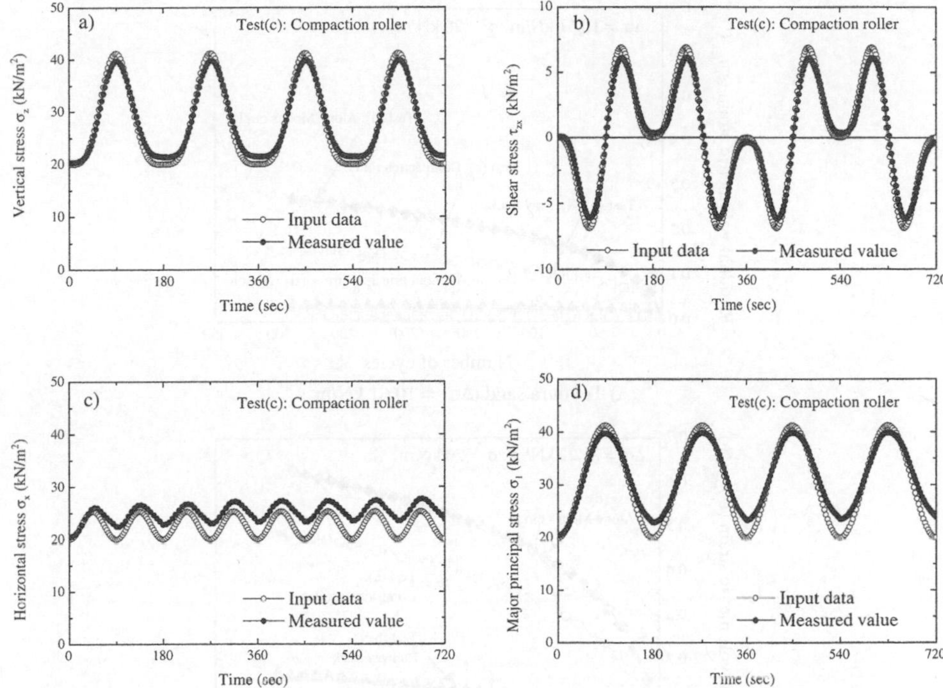

Figure 7. Typical input data and measured stresses, a) vertical stress σ_z, b) shear stress τ_{zx}, c) horizontal stress σ_x, d) Major principal stress σ_1 (test (c), $\Delta\sigma_1 = 21.22\,\text{kN/m}^2$).

principal stress axis rotation with constant despite that the magnitudes of all the principal stresses are kept constant during cyclic loading. Moreover, in test (d), the increasing rate of residual strain with cyclic loading did not decrease to a small value even after such a large number of cyclic loading as 300.

With both types of geomaterial, $(\varepsilon_v)_{res}$ was much smaller in test (b) (railway track) than in tests (c) & (b), while it was smallest in test (a) (without principal stress axis rotation), due to smaller principal stress axis rotation than tests (c) & (b). It is to be noted that the effects of continuous principal stress axis rotation on the residual strain by cyclic loading in test (b), simulating the stress conditions in the railway track, are not considerable, in particular with densely compacted well-graded gravelly soil (Fig. 8c). Yet, it is true that, when based on the results from cyclic loading tests without principal stress axis rotation, the residual strain by cyclic loading in the railway track is under-estimated to some extent.

4 CONCLUSIONS

A new special stress-strain test apparatus was developed to investigate the effect of principal stress axis rotation on the deformation characteristics of geomaterial

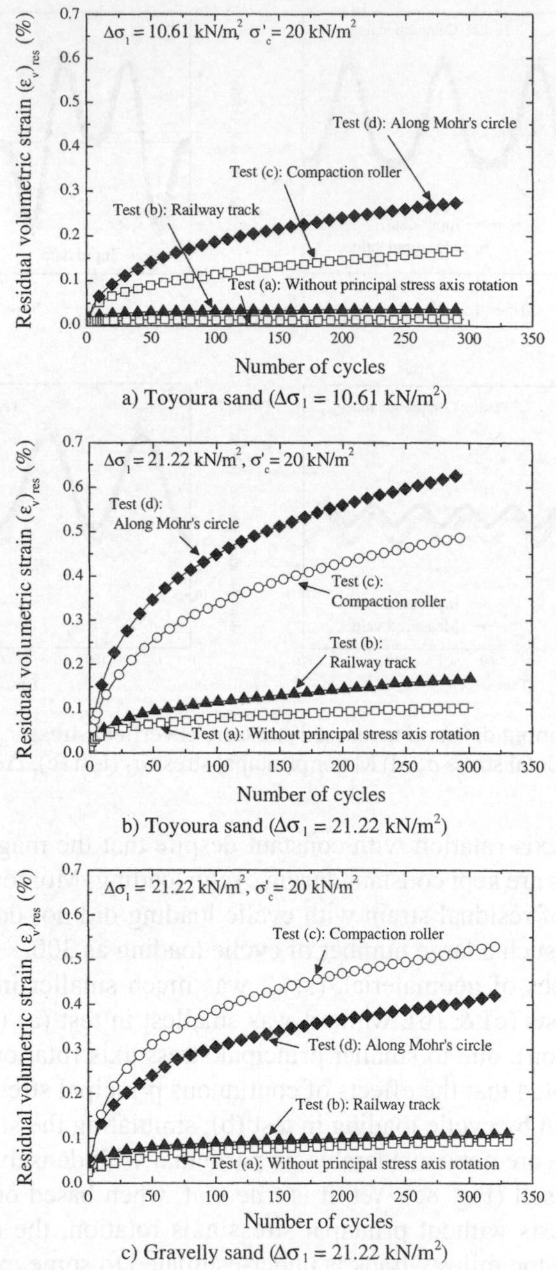

Figure 8. Residual volumetric strain under the cyclic loading tests with and without principal stress axis rotation.

subjected to cyclic loading. The apparatus can accommodate relatively material having large-diameter particles due to a relatively large specimen size (20 cm × 20 cm × 20 cm cubic). It was confirmed that the apparatus, developed based on the conventional simple shear test system, can control rather accurately the magnitudes and directions of the principal stresses under plane strain conditions. It was found however that it is necessary to introduce a feedback control system to improve the accuracy of stress control.

In tests on dense specimens on a fine uniform sand (Toyoura sand) and a well-graded gravelly soil, the residual strain by cyclic loading increased with an increase in the principal stress axis rotation. In particular, considerable residual strains took place by cyclic loading in test (d), in which the stress state changed along a fixed Mohr's circle of stress (thus with constant magnitudes of all the principal stresses). On the other hand, the residual strains by cyclic loading in test (b), simulating the stress conditions in the railway track, were much smaller and were not considerably larger than those in test (a) (without continuous principal stress axis rotation), in particular with densely compacted well-graded gravelly soil. Yet, the effects of continuous principal stress axis rotation should be taken into account when estimating residual strains by cyclic loading tests in the railway track.

REFERENCES

Hirakawa, D., Kawasaki, H., Tatsuoka, F. and Momoya,Y. 2002. Effects of loading conditions on the behaviour of railway track in the laboratory model tests, *Proc. 6th Int. Conf. on the Bearing Capacity of Roads, Railways and Airfields*, Vol. 2, pp.1295–1305, Lisbon: Balkema.

Hwang, S.K., Lee, S.H. and Choi, C.Y. 2001. Performance of the reinforced railroad roadbed of crushed stones under the simulated cyclic loading using multi purpose loading system, *Proceedings of the world Congress on Railway Research (WCRR2001)*, Germany.

Momoya,Y., Sekine, E. 2004. Reinforced roadbed deformation characteristics under moving wheel loads, *QR of RTRI*, Vol. 45, No. 3, pp.162–168.

Momoya,Y. and Sekine, E. 2005. Deformation characteristics of railway asphalt roadbed under a moving wheel load, *16th International Conference on Soil Mechanics and Geotechnical Engineering*.

Momoya, Y., Ando, K. and Horiike, T. 2002. Performance tests and basic design on solid bed track on asphalt pavement, *Proc. 6th Int. Conf. on the Bearing Capacity of Roads, Railways and Airfields,* Vol. 2, pp. 1307–1322, Lisbon: Balkema.

Towhata, I., Kawasaki, Y., Harada, N. and Sunaga, M. 1994. Contraction of soil subjected to traffic-type stress application, *Proc. of Inter. Sympo. on Pre-Failure Deformation Characteristics of Geomaterials*, pp. 305–310, Sapporo, Japan (Shibuya et al. eds.),.

Wong, R.K.S. and Arthur, J.R.F. (1985): Induced and inherent anisotropy in sand, *Géotechnique*, Vol. 35, No. 4, pp. 471–481.

8

The Effects of Shear Stress Reversal on the Accumulation of Plastic Strain in Granular Materials under Cyclic Loading

S.F. Brown
Nottingham Centre for Pavement Engineering, University of Nottingham, UK

ABSTRACT: Previous research has demonstrated that the shear reversal effect on stresses in a pavement caused by a moving wheel can have an important influence on the accumulation of plastic strains. In addition larger strains have been shown to develop under uni-directional wheel loading than under bi-directional loading. The shortcomings of both repeated load triaxial and simple shear tests in simulating the field condition lead to adoption of the hollow cylinder configuration. Simplified linear elastic analyses were used to demonstrate various stress paths generated in some typical pavement situations. Laboratory hollow cylinder test data are shown to illustrate the accumulation of shear and volumetric plastic strains in a Leighton Buzzard sand and a granulated (flaky) slate illustrating the importance of particle shape. The discussion and results are presented to stimulate the planning of future hollow cylinder testing to help develop an improved basis for the prediction of plastic strains in pavements and rail track.

1 INTRODUCTION

Over many years, the stress-strain relationships for soils and granular materials relevant to the foundations and lower layers of pavements and rail track have been studied using the repeated load triaxial test. Most effort has gone into understanding non-linear stress-resilient strain relationships but studies have also been conducted on the accumulation of plastic strain under repeated loading.

In the context of pavement design and structural evaluation, these two strain components are both of importance. Resilient characteristics are needed to carry out structural analysis of pavements in connection with design and the back-analysis of field loading tests. Plastic strains are relevant to the development of wheel track rutting in roads and settlement in rail track.

Brown (1996) presented a comprehensive review of the state of knowledge in this field. He pointed out the well-known shortcoming of triaxial stress conditions compared with the stress regime set up in a pavement from a rolling wheel load. This issue was first highlighted by Pell and Brown (1972) and relevant diagrams

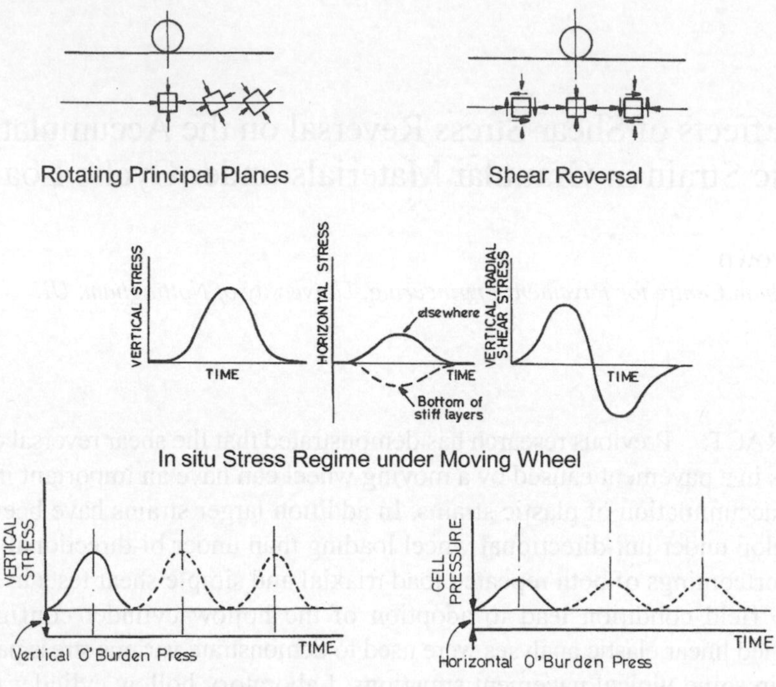

Figure 1. In situ stress regime induced by moving wheel load (after Pell and Brown, 1972).

from their paper are shown in Figure 1, which illustrates the key points. These are that the triaxial test can reproduce the normal stress regime but not in combination with the shear stress reversal. To deal with this issue, they pointed out that a direct shear test would be required.

This present paper describes some recent laboratory-based research on this matter against a background of developments over the past 35 years, during which period various attempts have been made to reproduce the complex pavement stress regime in laboratory element testing. This has led to application of the hollow cylinder configuration as a complex but realistic way of studying the relevant stress paths. Figure 2 shows that the stresses on an element of the cylinder wall are similar to those in the pavement.

2 PREVIOUS RESEARCH

Following the philosophy set out by Pell and Brown (1972), research at Nottingham initially focussed on development of a suitable simple shear test (SST) to reproduce the in-situ stress regime while, in parallel, high quality repeated load triaxial testing was being conducted (Brown, 1996). Ansell and Brown (1978) described the first

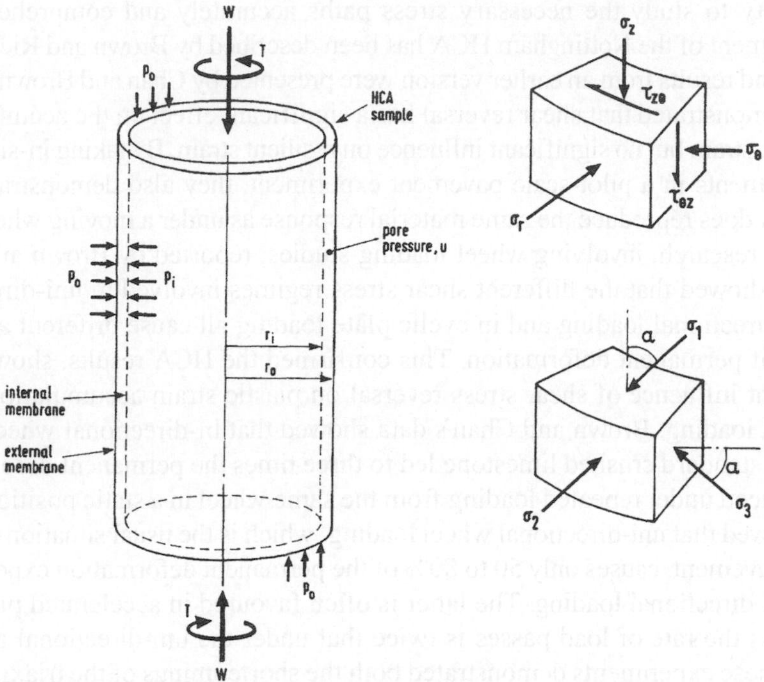

Figure 2. Schematic of the Hollow Cylinder Apparatus.

version of a pneumatically operated repeated load SST designed to test sand-sized dry crushed limestone based on the principles established for the Cambridge Simple Shear Test (Roscoe, 1953). They introduced a split top platen to accommodate differential dilation effects and focussed on quantifying the response of the middle third of the test specimen. Cyclic shear and constant normal stresses were applied. They clearly demonstrated that the application of shear stress reversal increased plastic volumetric strain and suggested, in a later paper (Brown and Ansell, 1980), that a horizontal component of vibration in rollers could improve the efficiency of compaction. They discovered, however, that the precise stress situation on the middle third of the specimen could not be adequately quantified.

Shaw and Brown (1986) described improvements to the Nottingham SST, which included application of cyclic normal stress and improved boundary stress measurements. By using a smaller particle size, they obtained more consistent results. However, they still experienced difficulties in making comparisons with data from triaxial tests to establish whether shear reversal was a significant factor. This was because of the continued difficulty in quantifying the stress conditions in terms of shear and normal invariants.

These difficulties led to abandonment of the simple shear test and adoption of the Hollow Cylinder Apparatus (HCA) which, though more complex, offered the possibility to study the necessary stress paths accurately and comprehensively. Development of the Nottingham HCA has been described by Brown and Richardson (2004) and results from an earlier version were presented by Chan and Brown (1994). They demonstrated that shear reversal has a significant effect on the accumulation of plastic strain but no significant influence on resilient strain. By taking in-situ strain measurements in a pilot-scale pavement experiment, they also demonstrated that the HCA does reproduce the same material response as under a moving wheel load.

Other research, involving wheel loading studies, reported by Brown and Chan (1996), showed that the different shear stress regimes involved in uni-directional and bi-directional loading and in cyclic plate loading all cause different accumulations of permanent deformation. This confirmed the HCA results, showing the important influence of shear stress reversal on plastic strain accumulation under repeated loading. Brown and Chan's data showed that bi-directional wheel tracking on a standard crushed limestone led to three times the permanent deformation experienced under repeated loading from the same wheel in a static position. They also showed that uni-directional wheel loading, which is the usual situation beneath a road pavement, causes only 50 to 80% of the permanent deformation experienced under bi-directional loading. The latter is often favoured in accelerated pavement testing as the rate of load passes is twice that under the uni-directional arrangement. These experiments demonstrated both the shortcomings of the triaxial stress configuration and, at a more detailed level, the differences in shear stress regimes between bi-directional and uni-directional wheel loading.

3 IN SITU STRESS CONDITIONS

While previously published diagrams, such as that in Figure 1, indicating the stress regime under a moving wheel load have been derived from values computed through pavement structural analysis, the basis for these was linear elastic theory. However, the response of soils and granular materials is known to be markedly non-linear (Brown, 1996). Brown and Pappin (1985) demonstrated that use of linear elasticity with a correctly selected equivalent single value of Young's modulus for the granular layer could be used to compute stresses, strains and deflections in other layers, but could lead to serious errors for positions within the granular layer itself. This is illustrated by Figure 3 from Brown and Pappin (1985), which shows the horizontal stresses in the granular layer of two pavement structures contrasting linear elastic layered system and non-linear elastic finite element solutions. The effect of including the non-linear characteristics and the self-weight stresses is very clear.

Notwithstanding the problems with use of linear elastic analysis, it was applied to two pavement systems for the purpose of providing some qualitative idea of the variation of shear stress on the vertical/horizontal planes at particular depths within the granular layer. The single value of Young's (resilient) modulus for the layer was based on the effective value derived by Brown and Pappin (1985) for a good qual-

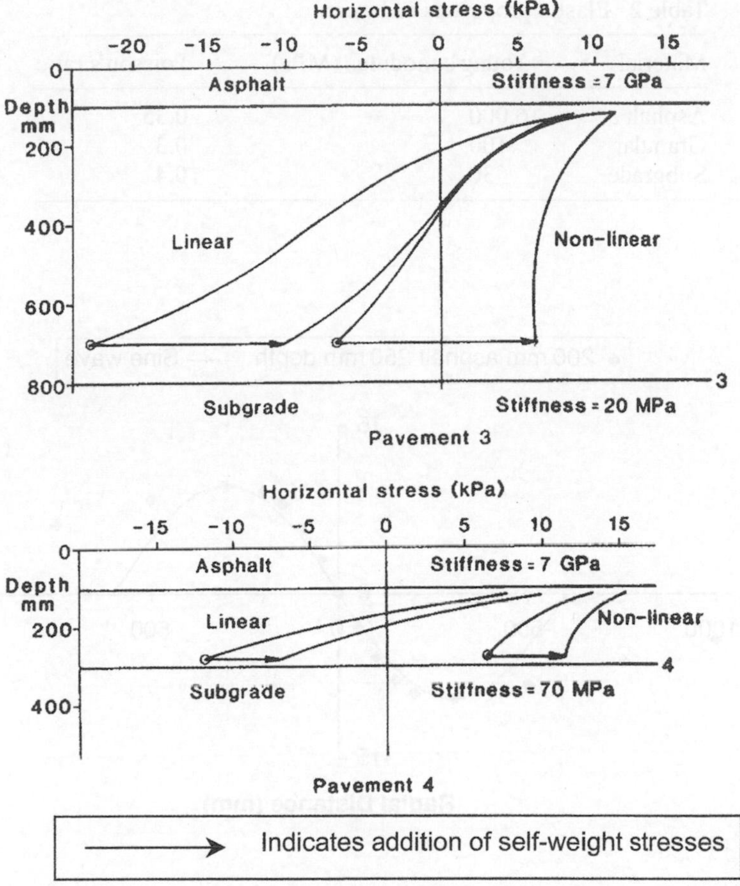

Figure 3. Computed Horizontal Stresses in Granular Layer (after Brown and Pappin, 1985).

Table 1. Details of Structures Analysed.

Structure	Layer thickness (mm)		Element depth (mm)
	Asphalt	Granular	
1	50	300	200
2	200	300	250

ity material. The linear elastic properties of each layer are shown in Table 2. A 40 kN single wheel load was applied over a uniformly distributed circular area of 160 mm radius at a pressure of 500 kPa. The stresses caused by the wheel load were calculated for elements at radial positions up to 1 m for the depths given in Table 1.

Figure 4 presents values for the shear stress on the vertical/radial planes, as shown for the element in Figure 5. Cases for the thick and thin asphalt construction are

Table 2. Elastic properties of layers.

Material	Young's modulus (MPa)	Poisson's ratio
Asphalt	6,000	0.35
Granular	100	0.3
Subgrade	30	0.4

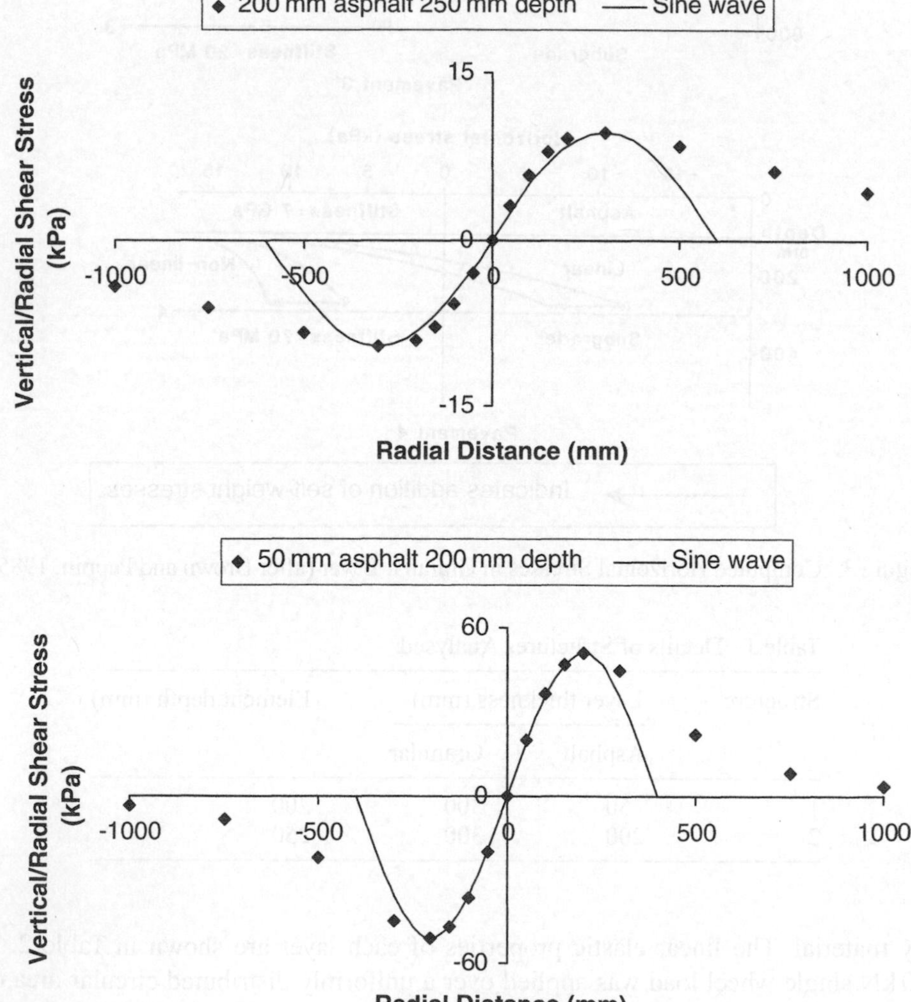

Figure 4. Variation of Vertical/Radial Shear Stress in the Granular Layer for Two Cases.

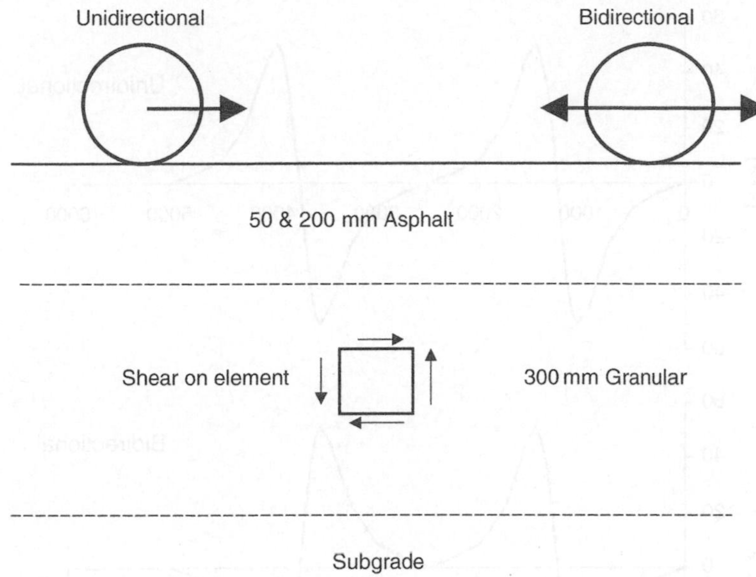

Figure 5. Definition of Vertical/Radial Shear on Element of Granular Layer.

given, from which it is apparent that the same pattern arises. Superimposed on each figure are sine wave forms arranged so that the peak values coincide with those for the computed stresses. This has been done since laboratory simulation tests generally use this form of loading. On the rise side of the stress pulse, a very good fit is apparent but the in situ stresses tail off more slowly than for the sine wave, although proper conclusions on this issue would need to be based on computations using non-linear models.

Some of this data was used to illustrate the different shear stress regimes that are caused by uni-directional wheel loading compared with bi-directional. Figure 6 shows the results, which are for the 50 mm asphalt construction and, again, relate to an element at the centre of the granular layer. This figure indicates why uni-directional loading is more damaging than bi-directional, since the latter involves two successive shear stress pulses in the same direction and explains the measurements reported by Brown and Chan (1996).

4 HOLLOW CYLINDER TESTING

A schematic of the Hollow Cylinder Apparatus (HCA) is shown in Figure 2. The equipment developed at Nottingham is shown in Figure 7. It was used for testing dry granular material having a nominal particle size of 0.5 mm.

The stress conditions in the walls of the HCA test specimen (Figure 2) result from the application of confining stress, in this case equally applied inside and outside,

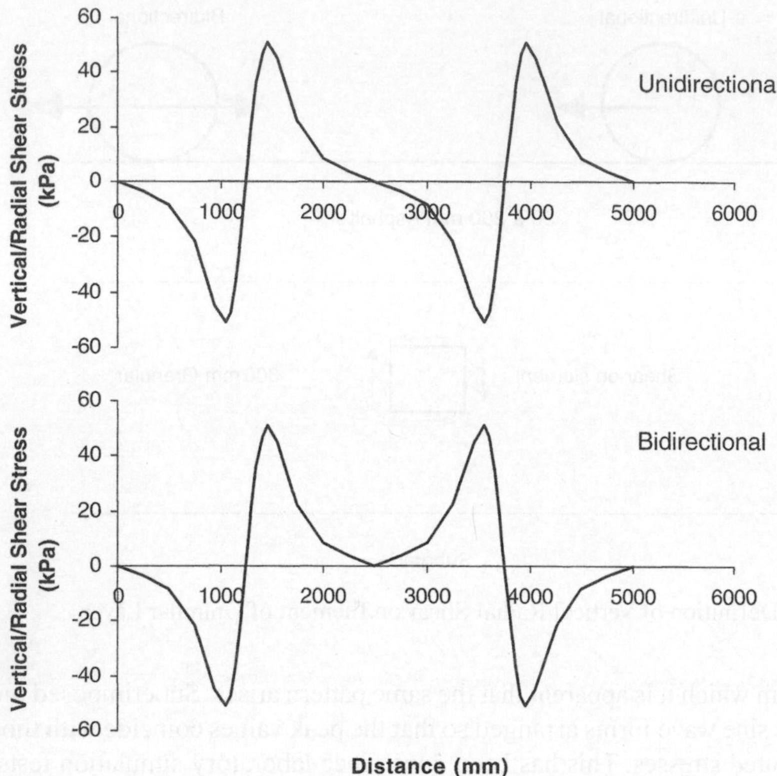

Figure 6. Computed Shear Stress Regimes for Uni- and Bi-directional Wheel Loading.

axial normal stress and torsion in the horizontal plane (Brown and Richardson, 2004).

The research conducted by Richardson (1999) involved a fundamental study of the effect of various stress paths on the monotonic and cyclic loading response of a uniform dry Leighton Buzzard sand and a flaky granulated slate of similar particle size. These materials were chosen to allow the influence of material anisotropy to be investigated. The parameters which were adopted to plan the experiments were as follows:

α = Angle between the direction of the major principal stress (σ_1) and the vertical,

η = Stress ratio (q/p′), where q is the deviator stress defined as:

$q = \sqrt{[(\sigma_1 - \sigma_2)^2 + (\sigma_2 - \sigma_3)^2 + (\sigma_3 - \sigma_1)^2]}/\sqrt{2}$

and p′ is the mean normal effective stress:

$p' = (\sigma_1 + \sigma_2 + \sigma_3)/3$

Since the materials were dry, total stress equalled effective stress.

The parameter η was used because previous studies (e.g. Pappin and Brown, 1980) had shown that shear strains were strongly related to it. The experiments

Figure 7. The Nottingham Hollow Cylinder Apparatus.

were planned so that the separate influences of α and η could be quantified under conditions of constant α and when principal stress rotation was applied (variable α).

Selected data from this study has been used to illustrate the effects of principal stress rotation on the accumulation of plastic strains under repeated loading. To do this realistically, the variations in both α and η for actual pavement structures should be quantified to determine the in situ relationship between these two parameters. Brown and Pappin (1985) showed that realistic values of stress ratio could only be determined from non-linear analysis. Once again, however, linear analysis was used to indicate the type of stress path in $\alpha - \eta$ space that is generated under a moving wheel load. This was done for one of the pavement structures considered by Brown and Pappin (1985). The characteristics of this pavement were slightly different from those specified in Tables 1 and 2. and are shown in Table 3.

The computed variation of α with radial distance from the load at a depth of 250 mm is shown in Figure 8 and that for η is in Figure 9, which also shows corrected values based on the finite element non-linear computations reported by

Table 3. Details of Pavement Structure No. 6 from Brown and Pappin (1985).

Asphalt			Granular			Subgrade	
Thickness (mm)	Young's Modulus (MPa)	Poisson's Ratio	Thickness (mm)	Young's Modulus (MPa)	Poisson's Ratio	Young's Modulus (MPa)	Poisson's Ratio
200	4,000	0.35	200	100	0.3	50	0.4

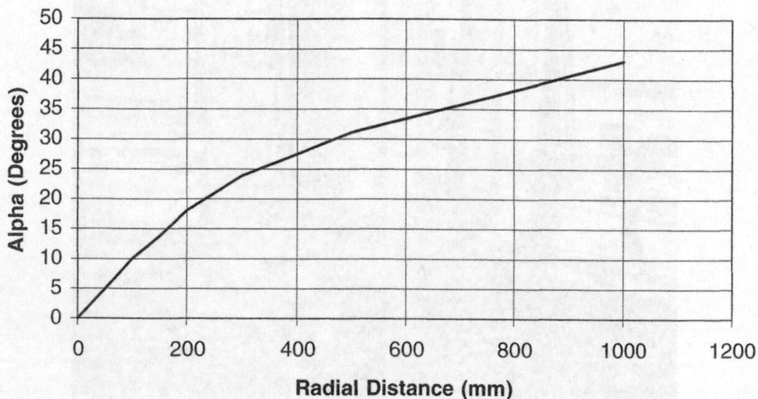

Figure 8. Variation of α with Radial Position at a depth of 250 mm for the Structure in Table 3.

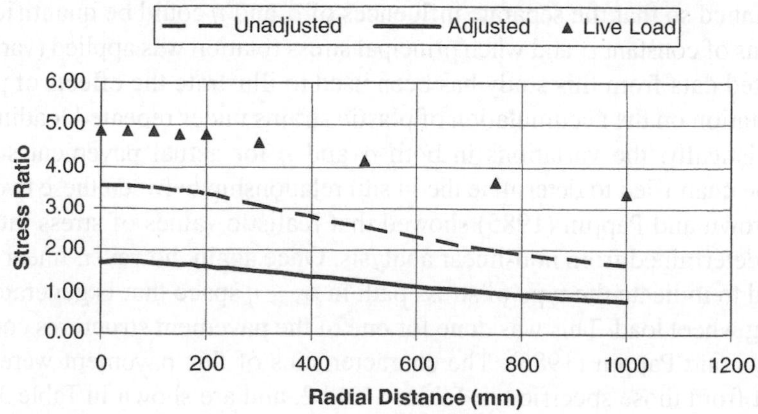

Figure 9. Variation of Stress Ratio (η) with Radial Position at a depth of 250 mm for various cases applied to the Structure in Table 3.

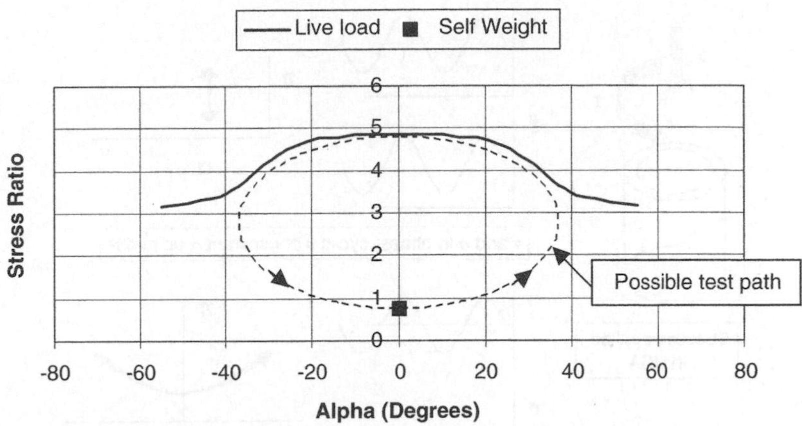

Figure 10. Relationship between α and η at a depth of 250 mm for the Structure in Table 3.

Brown and Pappin (1985). It is worth noting that the failure value of η for a good quality crushed rock sub-base is about 2 so only the non-linear solution is providing answers less than failure. This again clearly illustrates the qualitative shortcomings of the linear elastic assumption. Figure 9 also shows the variation of η based on the live load stresses alone. Laboratory simulation of field conditions needs to arrange for the stress path to combine the effects of self weight, representing the situation remote from the influence of the wheel load, and the superimposed stresses caused by the passing wheel. Unfortunately, values of α for non-linear analysis were not given by Brown and Pappin, so the relationship between α and η given in Figure 10 is based on the linear elastic results from Figures 8 and 9 for the live load only. The self weight condition is shown by the point at $\alpha = 0$, $\eta = 0.74$ for the depth considered (250 mm). This represents the starting point for the stress path and, as the wheel approaches, it will need to move from this point to join the $\alpha - \eta$ relationship. A possible simulation of this is superimposed in the figure. While this will be quantitatively inaccurate, it probably reflects the general shape of the relationship and is helpful as a basis for studying results from the Hollow Cylinder tests conducted by Richardson (1999).

The HCA controls allowed a wide range of stress paths to be applied under cyclic loading conditions and Figure 11 illustrates two possible combinations that were used in the experiments conducted by Richardson (1999). The top one was used to study the effects of cyclic η at constant α and the lower one involved cyclic variation of α about a mean value of zero, as occurs in the pavement. The actual data recorded for this second condition during an HCA test on Leighton Buzzard sand is shown in Figure 12. It shows the stresses which were applied and the resulting consistent repeated stress path in $\alpha - \eta$ space over a test of 10,000 cycles at 0.5 Hz. Figure 13 shows another example of HCA data involving cyclic variations

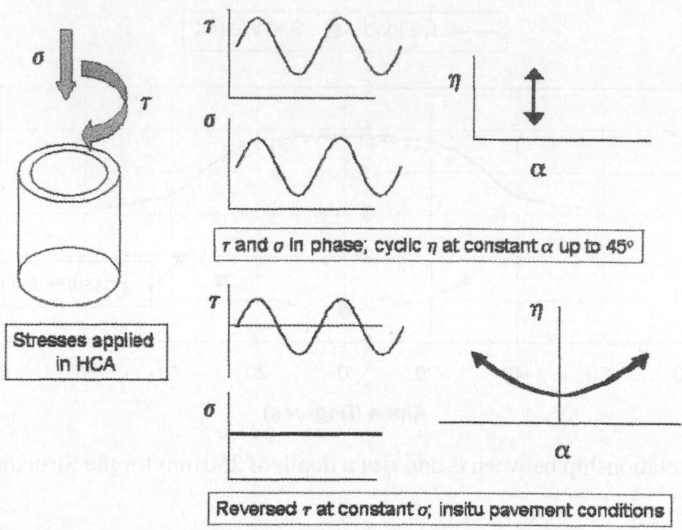

Figure 11. Examples of Stress Paths applied in the Hollow Cylinder Apparatus.

in both α and η. In this case the mean value of α is 45°, which does not simulate the field situation.

5 HCA RESULTS

The data presented in this section have been selected from the tests conducted by Richardson (1999). Although the control of the stress paths was good, the plastic strain results are rather difficult to interpret. It has not been possible, for instance, to identify the effects of principal stress rotation separately from other effects. However, some observations can be made which may be of interest and of assistance in planning future research.

Figures 14 and 15 show the accumulation of shear and volumetric plastic strains for tests on the Leighton Buzzard sand. These typical data indicate the relatively large strain accumulated in the first cycle of loading. This effect has been noted by others, particularly for single sized aggregate such as rail ballast (Shenton,1974). In order to obtain a better indication of the accumulation of strain under large numbers of stress cycles, Pappin (1979) proposed that the increase in strain after 100 cycles should be considered, thus eliminating the effects of the early cycles. He argued that these early effects were likely to be caused on site by construction traffic and were not relevant to the long-term performance of the pavement.

Richardson (1999) adopted the normalisation procedure of Pappin to present his results in a way which would allow comparison of accumulated plastic strains over

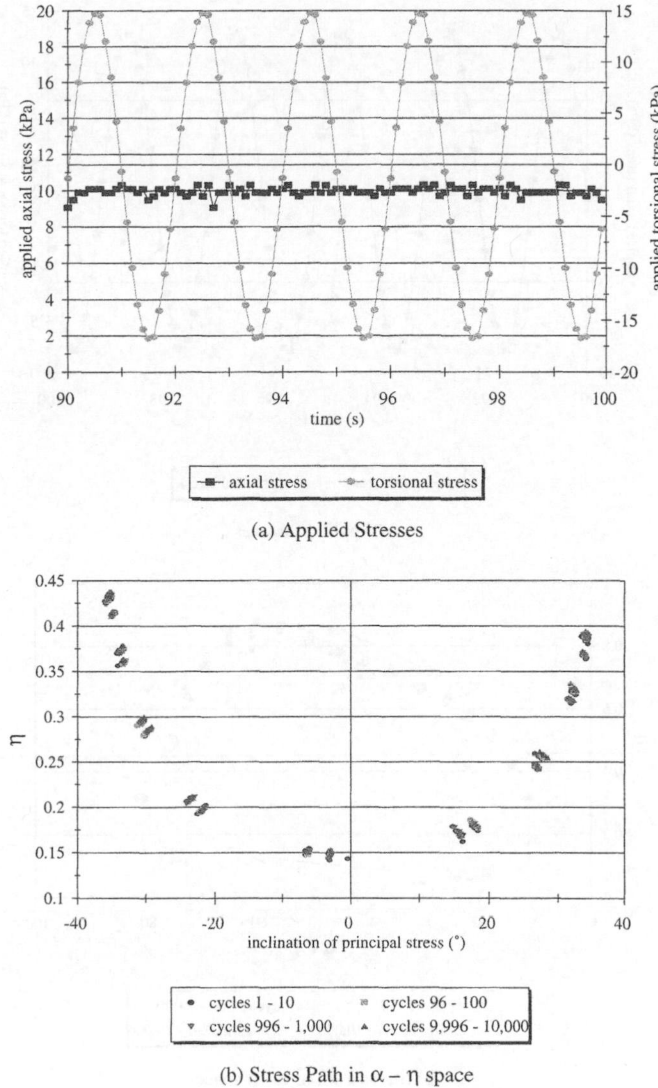

(a) Applied Stresses

(b) Stress Path in α – η space

Figure 12. Typical Recorded Data for an HCA test with Rotating Principal Planes (Mean $\alpha = 0$).

10,000 cycles of stress. He considered strain increases from the 10th cycle in the form of normalised strain plots with all results adjusted to a strain of 10 millistrain (1%) at cycle 10.

Figures 16 and 17 show the effect of α on the plastic shear and volumetric strains at a stress ratio, η, cycling from 5 to 70% of the static failure value. This parameter has been used in place of the actual value of η, since plastic strain accumulation

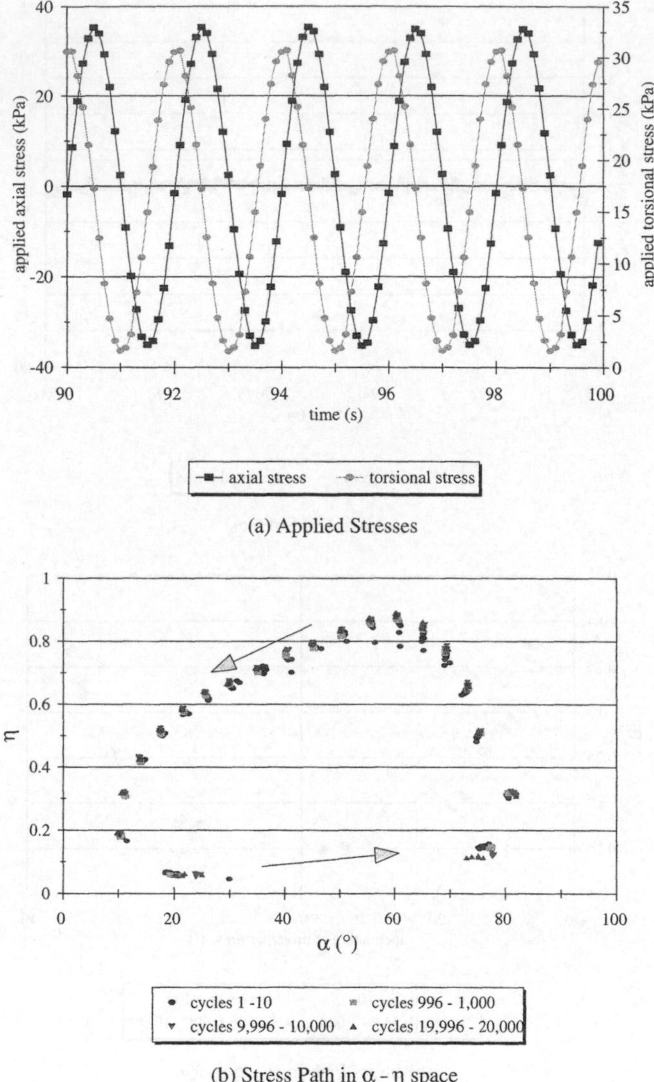

(a) Applied Stresses

(b) Stress Path in α - η space

Figure 13. Typical Recorded Data for an HCA test with Rotating Principal Planes (Mean $\alpha = 45^o$).

increases significantly as the peak stress ratio approaches the static failure value and it provides a better basis for comparing results. The data in Figures 16 and 17 are for the Leighton Buzzard sand and show that α has a significant effect on the strains but the qualitative relationship is not well defined. Higher shear strains developed for low values of α and dilation developed for intermediate values. Figures 18 and 19 show similar data for the Granulated Slate and demonstrate clearly

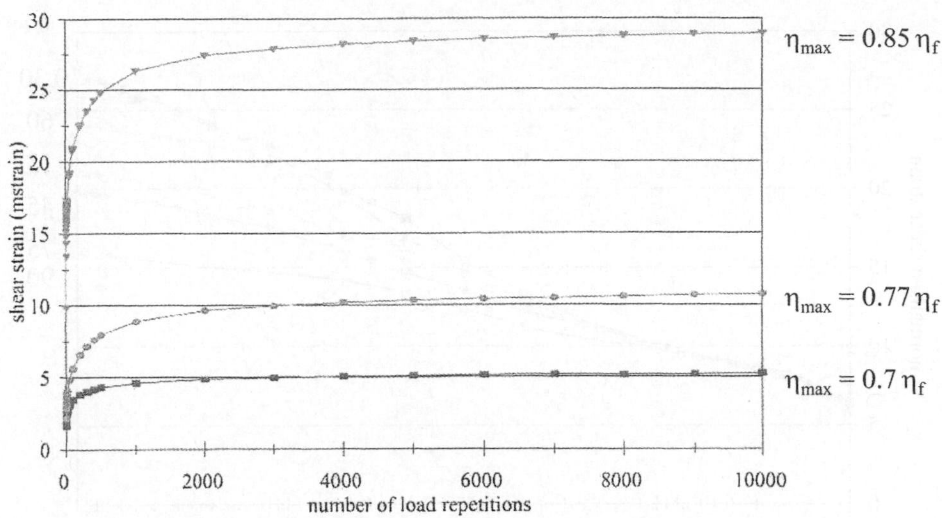

Figure 14. Shear Strain Development at α = 15° for Leighton Buzzard Sand.

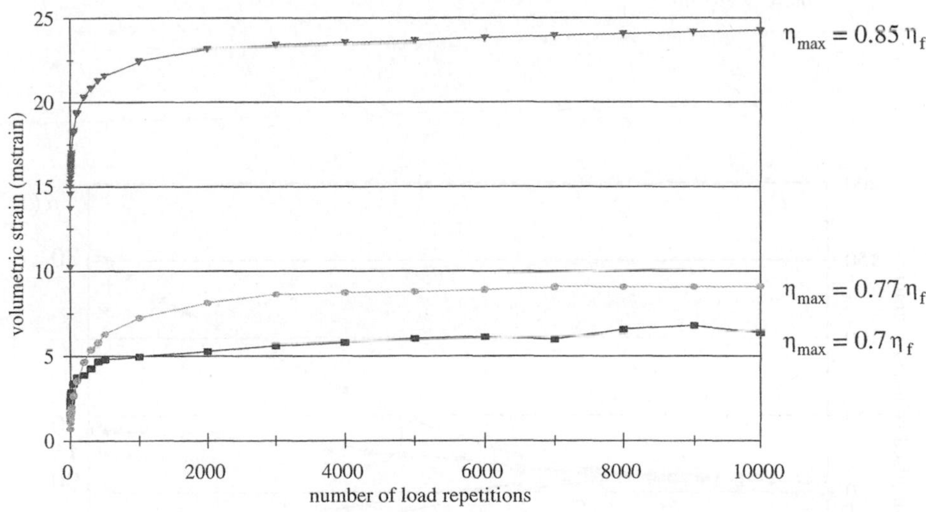

Figure 15. Volumetric Strain Development at α = 15° for Leighton Buzzard Sand.

different response for this anisotropic material which has flaky particles. α has little influence on shear strains but does influence volumetric strain with higher values associated with higher α but no evidence of dilation. These results demonstrate that particle shape does have a potentially major influence on the strain response

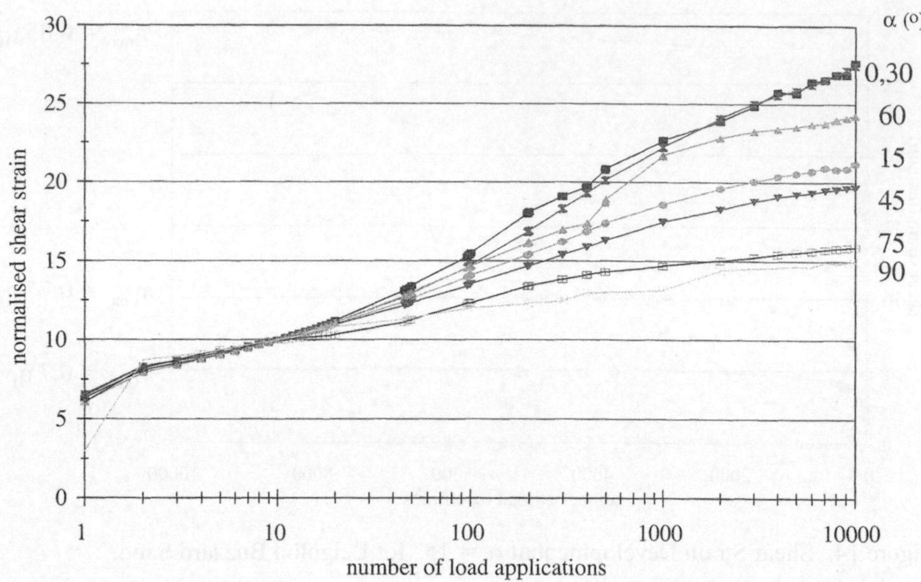

Figure 16. Influence of α on Accumulated Shear Strain in Leighton Buzzard Sand at $\eta_{max} = 0.7\,\eta_f$.

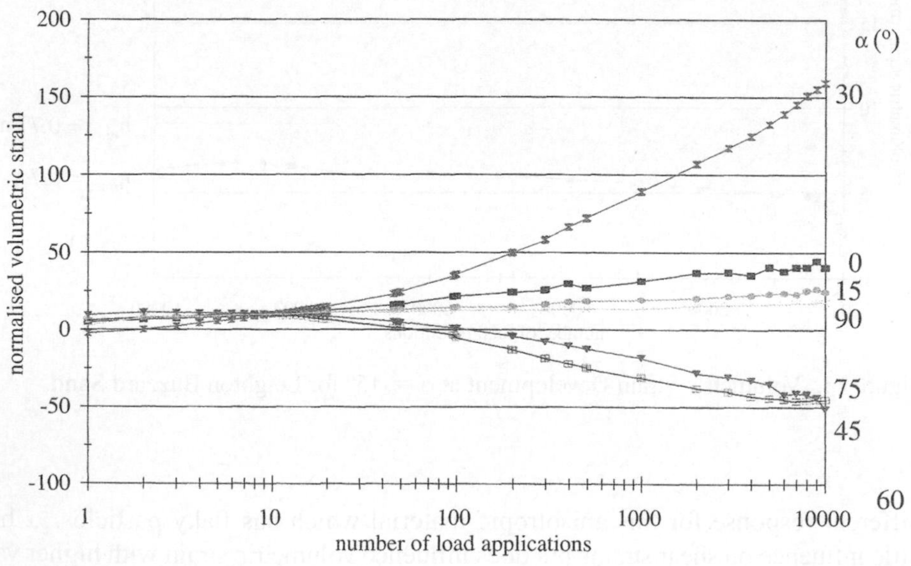

Figure 17. Influence of α on Accumulated Volumetric Strain in Leighton Buzzard Sand at $\eta_{max} = 0.7\eta_f$.

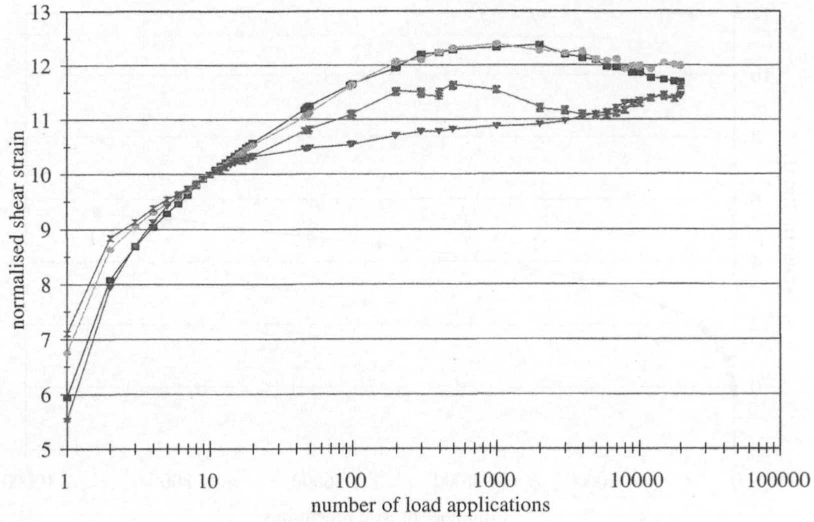

Figure 18. Influence of α on Accumulated Shear Strain in Granulated Slate at $\eta_{\max} = 0.7\,\eta_{\mathrm{f}}$ for $\alpha = 0$ to 67°.

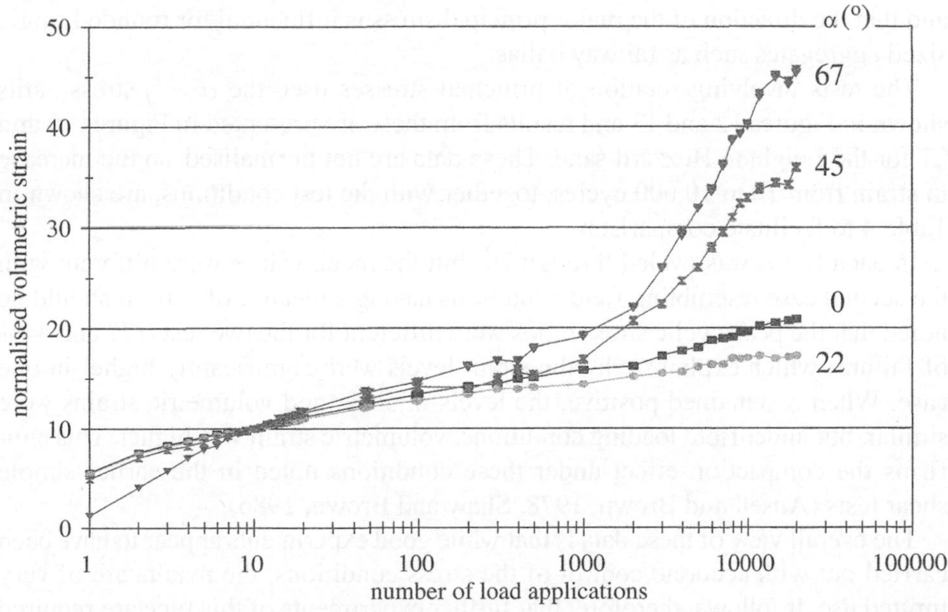

Figure 19. Influence of α on Accumulated Volumetric Strain in Granulated Slate at $\eta_{\max} = 0.7\,\eta_{\mathrm{f}}$.

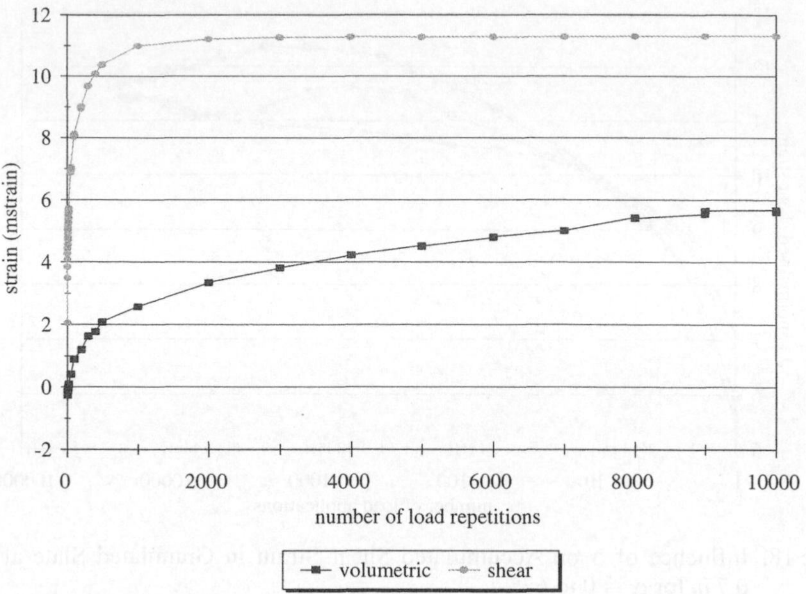

Figure 20. Accumulation of Strain in Leighton Buzzard with Cyclic α (10 to 80°) and Cyclic η ($\eta_{max} = 0.77\,\eta_f$).

and that the direction of the major principal stress is influential for rounded single sized aggregates such as railway ballast.

The tests involving rotation of principal stresses used the $\alpha - \eta$ stress paths shown in Figures 12 and 13 and results from these are presented in Figures 20 and 21 for the Leighton Buzzard sand. These data are not normalised, so the increase in strain from 10 to 10,000 cycles, together with the test conditions, are shown in Table 4 to facilitate comparison.

In each test α was cycled through 70° but the mean values were different with the second case resembling field conditions having a mean α of zero. It should be noted that the peak cyclic stress ratios were different for the two tests (77 and 44% of failure) which explains why the strain levels were significantly higher in one case. When α remained positive, the levels of shear and volumetric strains were similar, but under field loading conditions, volumetric strain was higher. This confirms the compaction effect under these conditions noted in the earlier simple shear tests (Ansell and Brown, 1978, Shaw and Brown, 1986).

The overall view of these data is that while good experiments appear to have been carried out with accurate control of the stress conditions, the results are of very limited use. It follows, therefore, that further experiments of this type are required with stress conditions appropriate to those that arise in road and rail construction and that a variety of material types need to be studied.

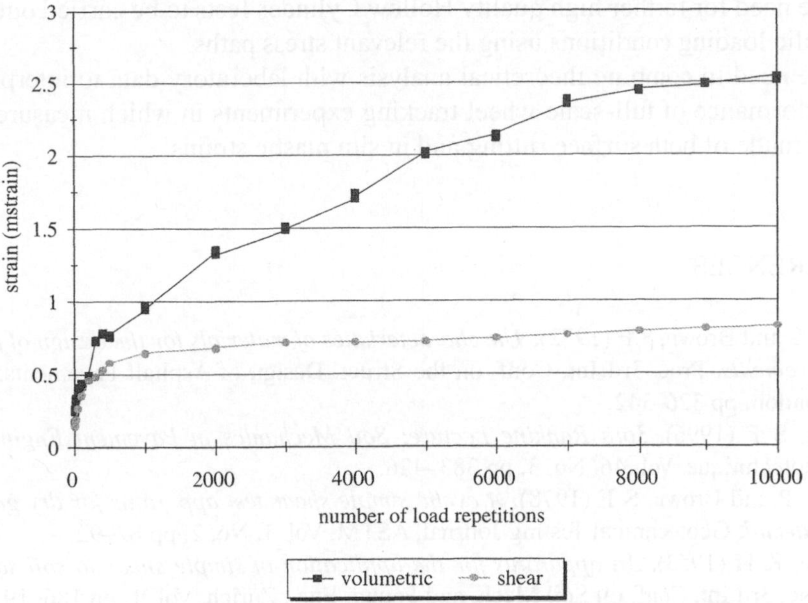

Figure 21. Accumulation of Strain in Leighton Buzzard with Cyclic α (−35 to +35°) and Cyclic η ($\eta_{max} = 0.44\,\eta_f$).

Table 4. Summary of Data from Tests on Leighton Buzzard Sand with Principal Stress Rotation.

Applied α (°)	η_{max}/η_f (%)	Strain from 10 to 10,000 cycles (%)		Strain after 10,000 cycles (%)	
		Shear	Volumetric	Shear	Volumetric
10 to 80	77	0.72	0.54	1.12	0.57
−35 to +35	44	0.06	0.23	0.08	0.25

6 CONCLUDING DISCUSSION

The ideas and results set out in this paper are intended to assist with discussion of the importance of principal stress rotation in the granular layers of pavements and rail track. The phenomenon is also of interest for asphalt and for the subgrade.

While no particularly new concepts have been presented and the new data are less than complete or conclusive, the discussion points to the following research requirements in this field:

1. The need for good quality non-linear analysis to be carried out on a range of pavement structures in order to quantify the stress conditions that need to be simulated in laboratory testing.

2. The need for further high quality Hollow Cylinder Tests to be carried out under cyclic loading conditions using the relevant stress paths.
3. The need to combine theoretical analysis with laboratory data to interpret the performance of full-scale wheel tracking experiments in which measurements are made of both surface rutting and in situ plastic strains.

REFERENCES

Pell, P S and Brown, S F (1972), *The characteristics of materials for the design of flexible pavements*, Proc. 3rd Int. Conf. on the Struct. Design of Asphalt Pavements,Vol. 1, London, pp 326–342.

Brown, S F (1996), *36th Rankine Lecture: Soil Mechanics in Pavement Engineering*, Géotechnique, Vol. 46, No. 3, pp 383–426.

Ansell, P and Brown, S F (1978), *A cyclic simple shear test apparatus for dry granular material*, Geotechnical Testing Journal, ASTM, Vol. 1, No. 2, pp 82–92.

Roscoe, K H (1953), *An apparatus for the application of simple shear to soil samples*, Proc. 3rd Int. Conf. on Soil Mech. and Found. Eng., Zurich, Vol. 1, pp 186–191.

Brown, S F and Ansell, P (1980), *The influence of repeated shear reversal on the compaction of granular material*, Proc. Int. Conference on Compaction, Vol. 1, Paris, pp 25–27.

Shaw, P and Brown, S F (1986), *Cyclic simple shear testing of granular materials*, Geot. Testing Journal, ASTM, Vol. 9, No. 4, pp 213–220.

Brown, S F and Richardson, R (2004), *A hollow cylinder apparatus to study the cyclic loading behaviour of dry granular material*, Advances in Geotechnical Engineering, Vol. 1, Thomas Telford, London, pp 369–380.

Chan, F W K and Brown, S F (1994), *Significance of principal stress rotation in pavements*, Proc. 13th Int. Conf. on Soil Mech. and Found. Eng., Vol. 4, Delhi, pp 1823–1826.

Brown, S F and Chan, F W K (1996), *Reduced rutting in unbound granular pavement layers through improved grading design*, Proc. Inst. of Civil Engineers Transport, Vol. 117, pp 40–49.

Brown, S F and Pappin, J W (1985), *Modelling of granular materials in pavements*, Transp. Res. Rec. No. 1022, Transp. Research Board, Washington, DC, pp 45–51.

Richardson, I R (1999), *The stress-strain behaviour of dry granular material subjected to repeated loading in a hollow cylinder apparatus*, PhD Thesis, Univ. of Nottingham.

Pappin, J W and Brown, S F (1980), *Resilient stress-strain behaviour of a crushed rock*, Proc. Int. Symp. on soils under cyclic and transient loading, Vol. 1, Swansea, pp 169–177.

Shenton, M J (1974), *Deformation of railway ballast under repeated loading triaxial tests*, Soil Mech. section, British Rail Research Dept. Derby.

Pappin, J W (1979), *Characteristics of granular material for pavement analysis*, PhD Thesis, Univ. of Nottingham.

3. Earth Structures in Pavement and Railway Construction – Promoting the Use of Processed Materials and Continuous Compaction Control

9

Roller-integrated Continuous Compaction Control (CCC) Technical Contractual Provisions & Recommendations

D. Adam
Vienna University of Technology & Consulting Engineer, Austria

ABSTRACT: In the scope of ISSMGE TC3 activities (Geotechnics of pavements) quality control methods based on roller-integrated continuous compaction control (CCC) were standardized and technical contractual provisions were established. In this paper the specifications and recommendations are introduced and the unabridged text is presented. CCC is an innovative technique in order to control and check the quality of compacted layers already during the compaction process performed by vibratory rollers. The motion behavior of the dynamically excited roller drum interacting with the ground is measured and data are immediately available to evaluate the properties of the ground visualized for the roller operator and documented for quality assurance.

INTRODUCTION

The quality of roads, highways, motorways, rail tracks, airfields, earth dams, waste disposal facilities, foundations of structures and buildings, etc. depends highly on the degree of compaction of filled layers consisting of different kinds of materials, e.g. soil, granular material, artificial powders, fly ashes and grain mixtures, unbound and bound material. Thus, both compaction method and compaction equipment have to be selected carefully taking into consideration the used material suitable for the prevailing purpose. Compaction process should be optimized in order to achieve sufficient compaction and uniform bearing and settlement conditions.

If compaction control can be included in the compaction process, time can be saved and cost reduced. Furthermore, a high-leveled quality management requires continuous control all over the compacted area, which can only be achieved economically by roller-integrated methods. The roller-integrated continuous compaction control (CCC) provides relative values representing the developing of the material stiffness all over the compacted area. These values have to be calibrated in order to relate them to conventional values (deformation modulus of static load plate test) given in contractual provisions and standards.

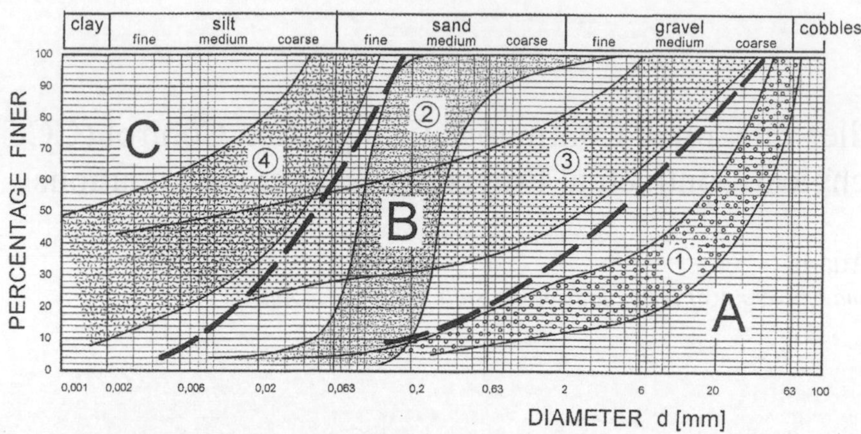

Figure 1. Application of CCC to different materials: A – excellent, B – good, C – moderate.

CCC can be applied to all kinds of unbound material. Coarse, widely grained material (gravel, sandy gravel) is best suited for CCC application (Figure 1, A). Single size fraction sands (Loess) tend to surface near re-loosening due to dynamic loads and achieve only a low maximum density compared to other soil types. In widely grained soil (clayey, silty and sandy gravel) containing a high amount of fines (ca. 30–50%) the water content influences the compaction behavior noticeably. The higher the moisture content the more water is trapped in the voids of the low permeable material. Consequently, pore water pressures reduce the compactibility more and more (Figure 1, B) and thus, influence CCC values. Fine grained soil and artificial material (e.g. fly ash) can hardly be compacted due to the low water and air permeability. Pore water and air produce excessive pore pressure during the compaction process. Sufficient compaction can only be gained by kneading the material to relieve the pore pressures (Figure 1, C).

Dynamic rollers make use of a vibrating or oscillating mechanism, which consists of one or more rotating eccentric weights. During dynamic compaction a combination of dynamic and static loads occurs. The dynamically excited drum delivers a rapid succession of impacts to the underlying surface from where the compressive and shear waves are transmitted through the material to set the particles in motion. This eliminates periodically the internal friction and facilitates the rearrangement of the particles into positions that result in a low void ratio and a high density. Furthermore, the increase in the number of contact points and planes between the grains leads to higher stability, stiffness, and lower long-term settlement behavior.

Nowadays, vibratory rollers and VARIO rollers can be used for CCC. Thus, their mechanisms are presented briefly in the following.

The drum of a vibratory roller is excited by a rotating eccentric mass which is shafted on the drum axis (Figure 2). The rotating mass sets the drum in a circular

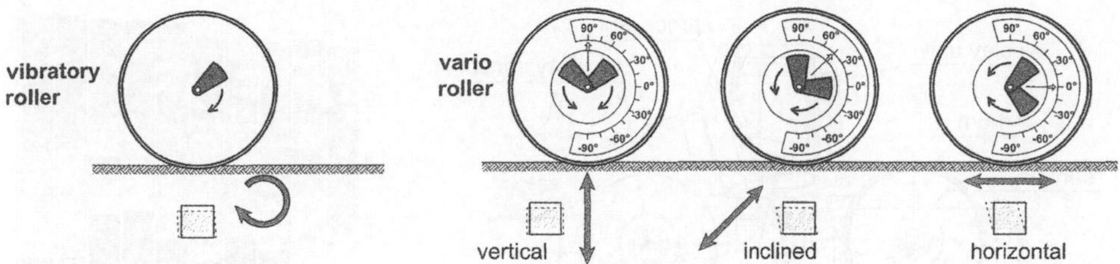

Figure 2: Dynamic roller and different kinds of excitation: vibratory roller and VARIO roller.

translatory motion, i.e. the direction of the resulting force is corresponding with the position of the eccentric. Compaction is achieved mainly by transmitted compression waves in combination with the effective static drum load. Consequently, the maximum resulting compaction force is supposed to be almost vertical and in fact it is only a little inclined. The vibration of the roller drum changes in dependence of the soil response. Numerous investigations have revealed that the drum of a vibratory roller operates in different conditions depending on roller and soil parameters. Five operating modes specified in the technical provisions can occur (Adam 1996).

In a VARIO roller two counter-rotating exciter masses, which are concentrically shafted on the axis of the drum, produce a directed vibration. The direction of excitation can be adjusted by turning the complete exciter unit (Figure 2) in order to optimize the compaction effect for the respective soil type. If the exciter direction is (almost) vertical or inclined, the compaction effect of a VARIO roller can be compared with that of a vibratory roller. However, if the exciter direction is horizontal, VARIO rollers compact soil like an oscillatory roller. Thus, a VARIO roller can be used both for dynamic compression compaction, for dynamic shear compaction, and a combination of these two possibilities, depending only on the adjustable force direction. Thus, VARIO rollers can be employed universally for each soil type, the respective optimum direction can be found by basic site investigations (Kopf 1999).

Based on the findings relating to the operation modes of different dynamic rollers, the company BOMAG developed the first automatically controlled so called VARIO CONTROL roller. The Swiss company AMMANN developed the auto-controlled roller ACE in connection with a roller-integrated control system providing dynamic compaction values independent from roller parameters. Exemplary, in VARIO CONTROL rollers the direction of excitation (vibrations can be directed infinitely from the vertical to the horizontal direction) is controlled automatically by using defined control criteria, which allow an optimized compaction process and, consequently, a highly uniform compaction. Such feedback controlled rollers can only be used as CCC-measuring tools, if the machine parameters are kept constantly for a measurement pass.

The roller-integrated continuous compaction control (CCC) is based on the measurement of the dynamic interaction between dynamic rollers and soil (Adam

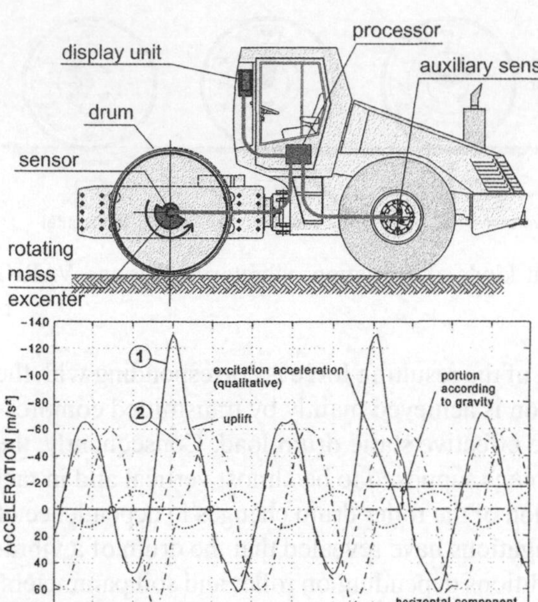

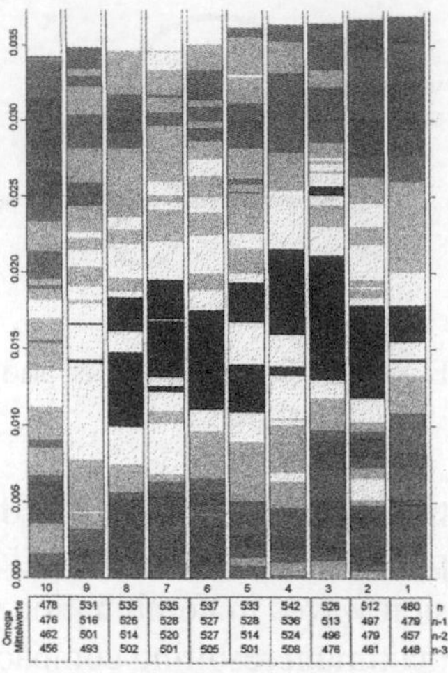

Figure 3. CCC-principle: CCC-components; from measured drum acceleration CCC-values are derived, and displayed area plot comprises the distribution of CCC-values.

1996). The motion behavior of different dynamically excited roller drums changes in dependence of the soil response. This fact is used to determine the stiffness of the ground. Accordingly, the drum of the dynamic roller is used as a measuring tool; its motion behavior is recorded, analyzed in a processor unit where a dynamic compaction value is calculated, and visualized on a dial or on a display unit where data can also be stored. Furthermore, an auxiliary sensor is necessary to determine the location of the roller or the localization is GPS-based. Control data are already available during the compaction process and all over the compacted area (Figure 3).

Four recording systems are available for vibratory rollers and VARIO rollers with vertical or any inclined excitation direction (except horizontal direction). All systems consist of a sensor containing one or two accelerometers attached to the bearing of the roller drum, a processor unit and a display to visualize the measured values. The sensor continuously records the acceleration of the drum. The time history of the acceleration signal is analyzed in the processor unit in order to determine dynamic compaction values with regard to specified roller parameters.

Table 1 gives a review of the recording systems of CCC. All defined CCC-values have proven suitable for roller-integrated checking of the actual compaction state. Nevertheless, it is essential to take into consideration the operating condition of the roller drum. Figure 4 shows the progress of CCC-values depending on soil stiffness and relative vertical drum amplitude.

Table 1. Established CCC-systems, CCC-values and their definitions; producers.

CCC-system	CCC-value	Definition of CCC-value	Manufacturer
Compactometer	**CMV** [ ]	acceleration amplitude ratio (first harmonic div. by excitation frequency amplitude) – *frequency domain*	Geodynamik, Sweden
Terrameter	**OMEGA** [Nm]	energy transferred to soil (considering soil contact force displacement relationship of 2 excitation cycles) – *time domain*	Bomag, Germany
Terrameter	$\mathbf{E_{vib}}$ [MN/m^2]	dynamic elasticity modulus of soil beneath drum (inclination of soil contact force displacement relationship during loading) – *time domain*	Bomag, Germany
ACE	$\mathbf{k_B}$ [N/m]	spring stiffness of soil beneath drum (derived from soil contact force displacement relationship at maximum drum deflection) – *time domain*	Ammann, Switzerland

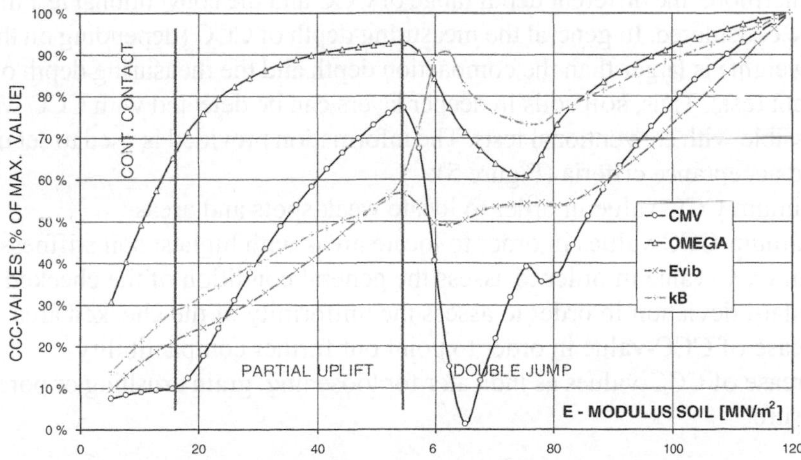

Figure 4. Relative CCC-values depending on soil stiffness.

The dynamic compaction values are relative values having a clear physical background (see Table 1). If the data shall be compared with common conventional values like the deformation modulus of the static or dynamic load plate test calibrations have to be performed. There are several possibilities to select the spots where

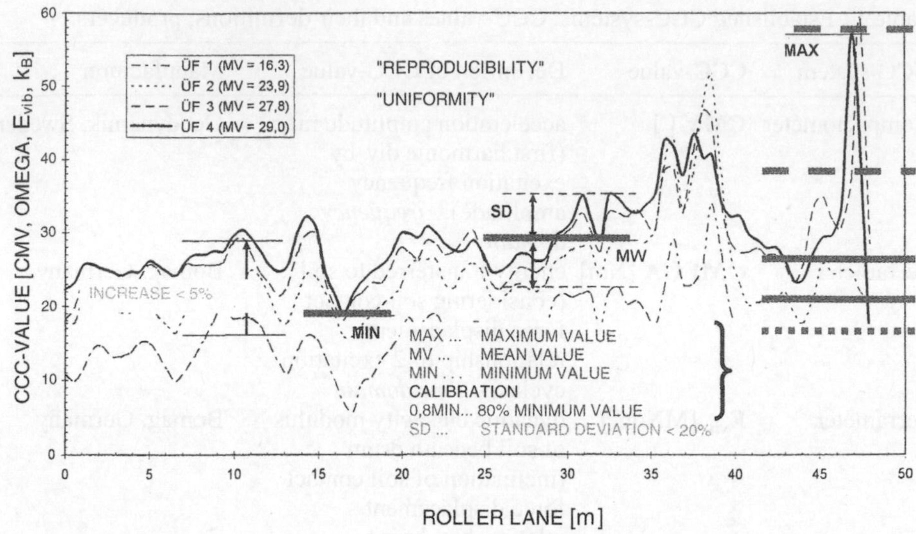

Figure 5. Progress of CCC-data and control criteria for standardized application; defined limits.

conventional tests can be carried out. Best correlation can be achieved, if the spots are selected by means of CCC-results. Spots with high, mean, and low dynamic compaction values indicate a wide range of soil properties.

Furthermore, the different depth range of CCC and the conventional test methods must be considered. In general the measuring depth of CCC (depending on the total roller weight) is larger than the compaction depth and the measuring depth of common spot tests. Thus, soft soils in deeper layers can be detected with CCC which is not possible with conventional tests. The information provided is used to set up control and acceptance criteria (Figure 5):

- Minimum CCC-value in order to locate weak spots and areas
- Maximum CCC-values in order to locate areas with highest soil stiffness
- Mean CCC-value in order to assess the general condition of the checked area
- Standard deviation in order to assess the uniformity of the checked area
- Increase of CCC-value in order to point out further compactibility
- Decrease of CCC-values as indicator for loosening, grain crushing or pore water pressure

1 APPLICATION

This technical provision serves to prove continuously the compaction of the different layers and planes (levels) in road construction by means of roller-integrated measuring systems. Its application is also recommended for railways, airfields, embankments, fill dams, landfills, soil replacement and all kinds of earth structures and subgrade compaction of flat foundations.

This roller-integrated dynamic continuous compaction control (CCC) makes it possible to check and document the values demanded in the relevant contractual provisions.

2 DEFINITIONS

Measuring roller
Measuring rollers are all vibratory rollers equipped with a compaction measuring system.

(Double) jump operation
The drum of the roller used is in (double) jump operation when the amplitude of the drum acceleration at half basic frequency of the exciter acceleration is different from zero (see appendix, item 1).

Track plot
A track plot is an EDP printout of the measured values along a roller track.

Area plot
An area plot is an EDP printout of the measured values of the particular area tested.

Measuring depth
The measuring depth is the depth in the soil under the drum up to where soil properties have an influence on the dynamically measured values.

Compaction depth
The compaction depth is the depth in the soil up to where the roller used on the specific site achieves a compaction effect.

3 PRINCIPLE OF ROLLER-INTEGRATED CONTINUOUS COMPACTION CONTROL WITH VIBRATORY ROLLERS

Roller-integrated continuous compaction control (CCC) is based on the dynamic interaction between the excited drum of a vibratory roller and the soil or other granular material that has to be compacted. The dynamic measuring value determined from the movement behaviour of the drum must be physically clearly defined.

Vibratory rollers are characterised by a drum that is excited by an eccentric mass rotating at constant speed. However, the excitation can also be generated by two counter-rotating eccentric masses, whereby these masses must be arranged in such a way during measuring passes, that the drum is exclusively vertically excited.

The method is suitable for all soils and grain mixtures as well as for recycling materials which are dynamically compactable. The measurements are performed during the compaction process, so that the increase in compaction (compaction effect) becomes continuously evident and weak points can be detected already at an early stage.

In principle, it is also possible to subject soils, which were statically compacted, subsequently to dynamic testing.

4 APPLICATION OF ROLLER-INTEGRATED CCC

Roller integrated measuring systems are available at a high-tech quality and they are economically usable:

- In the base of embankments and on the surface of improved sub-grade (soil replacement) as well as from 1 m below substructure level.
- In lower and upper unbound wearing courses.
- On cement-stabilised wearing courses during the compaction process and immediately after.
- For the generation of micro cracks in hydraulically bonded wearing courses.

5 PREREQUISITES AND REQUIREMENTS

5.1 Soil (and other granular material)

5.1.1 *Soil types*

Roller-integrated measuring systems can be used for all types of soil and for all unbound grain mixes.

Should the maximum grain size exceed 120 mm, particular attention must be paid to the evenness of the soil level. If a fine grain proportion <0.006 mm exceeds 15 mass%, special attention must be given to the water content.

5.1.2 *Requirements for the surface*

- The soil surface should be even, e.g. no truck ruts; the drum must have ground contact over its entire width.
- The surface should be rather homogeneous, i.e. no strongly varying material such as big stones, accumulation of fine particles or strongly deviating water content. Otherwise the interpretation of the measured results is difficult.

5.2 Rollers

5.2.1 *Measuring rollers*

- Single drum vibratory rollers (vibratory rollers driven by rubber wheels) with smooth drum, which may also be driven, provide the best results with respect to constant travel speed. Furthermore, advantageous are their higher moveability and problem-free use on slopes and loose surfaces. It is also possible to use a padfoot roller instead of the smooth drum roller.

- Towed rollers and combination rollers may also be used as measuring rollers as long as the speed is kept constant and documented, and the allocation of the measured values to the particular place in the field is guaranteed.
- Tandem vibratory rollers with two smooth drums are less suitable. Under particular subgrade and terrain conditions (e.g. slopes) these rollers may sometimes undergo a clear "slip" of the driven drums. The travel speed becomes sometimes uncontrollable then and the measured values cannot be assigned to the correct place in the field.
- The CCC-method is not applicable with static and vibratory rollers in static operation.

5.2.2 *Requirements for the measuring rollers*

- The exciter frequency must be kept constantly within a certain range (see item 6.3.2).
- A constant travel speed must be kept constantly within a certain range (see item 6.3.4)
- The vibration behaviour of the drum must be reproducible. Possible periodicities due to roller influences (e.g. worn out bearings, unbalanced drum) have to be rectified by appropriate repair; otherwise such rollers must not be used as measuring rollers.

5.3 Measuring system

5.3.1 *Structure of the measuring system*

The measuring system consists of the following system units which are linked to each other by appropriate cables:

- transducer (measurement of the drum acceleration)
- processor unit (calculation of the dynamic measuring value)
- display unit (display of the actual measuring value for the roller driver, e.g. screen, analogue clock, paper strip)
- documentation unit (storage of measured data and co-ordinated stationing of the measured values secured against manipulation, so that they are available for a subsequent evaluation)
- transducer for position-wise assignment for the dynamic measuring values within a roller track or positioning system

For assembly and operation, the instructions given by the manufacturer of the measuring system are to be observed.

5.3.2 *Requirements for the measuring system*

- Measurement and documentation of travel speed, frequency, selected theoretical amplitude and the dynamic measuring values as well as their position-wise assignment and the occurrence of jump operation must be guaranteed in such a way that there is no manipulation possible.

- Installation of the CCC measuring system on conventional rollers must be possible.
- The measuring system must be EDP-compatible and enable data storage and documentation on the spot as well as a subsequent evaluation.
- The measured values must be assignable by means of co-ordinates; therefore, an exact adherence to the roller track is required.
- The input of the required roller parameters must be possible and their compliance be documented.
- The measuring system must enable a clear presentation of the required values and must display them directly to the roller driver.
- The travel speed must be recorded and documented during the measurement.
- Jump operation is to be indicated and documented also with its entire position.
- Structure of the measuring system.

5.4 Reproducibility

Series of measurements which were determined subsequently (directly after each other) on the same roller track must exhibit similar courses of development, i.e. significant measuring values must widely re-occur at the same positions during successive passes.

Otherwise, the reproducibility for the unit "soil-roller – measuring system" with specified amplitude, frequency and travel speed must be checked.

A possibility for reviewing the reproducibility is shown in the appendix (item 1).

5.4.1 *Different measuring rollers with the same measuring system*

It is allowable to use different measuring rollers with the same measuring systems. Within a specific operating condition, the courses of measuring values are similar, but the level of measured values is different.

5.4.2 *Same measuring roller type with different measuring systems*

It is allowable to use different measuring systems on the same type of roller. Within a specific operation condition (see item 6.1) the courses of measuring values are similar, but the level of measuring values is different.

5.5 Personnel requirements

All persons involved in the roller-integrated continuous compaction control must be familiar with CCC in accordance with their fields of responsibility.

6 INFLUENCES ON THE MEASURED VALUES

6.1 Operation conditions of the drum

Different operating conditions of the drum have a significant effect on the level of dynamic measuring values. The dynamic measuring values have to be evaluated separately – on the one hand *commonly* for "contact" and "partial uplift", and on the other hand for "(double) jump" – whereby the latter should be characterized as "(double) jump operation".

(Double) jump operation should be avoided as far as possible. This is primarily achieved by the choice of a smaller amplitude and/or a higher roller speed.

The appendix (item 1) contains the definition of the operating conditions and explanatory notes for better understanding.

6.2 Soil (and other granular material)

6.2.1 *Layer structure*

Layered soils influence the level of the dynamic measuring values and possibly the motion behaviour of the drum (operating condition) if the interface(s) of the layers lie(s) within the measuring depth.

Weak points in lower areas have an influence on the course of the dynamic measuring values.

6.2.2 *Soil type*

Grain size and grain distribution influence the dynamic measuring values, their level and development in the case of subsequent roller passes, and possibly also the motion behaviour of the drum (operating condition).

6.2.3 *Water content*

In cases where proper compaction is not possible due to a too high water content of the material to be compacted, a lower level of the dynamic measuring values is registered, which does not increase, but decreases over several passes.

Softened surface (after strong rain) also exhibits a decrease of the level of measured values.

6.2.4 *Inhomogeneities near the surface*

The layer or zone to be controlled should consist as far as possible of homogeneous material, since locally strongly varying contact conditions of the drum (e.g. sand on one side and stones on the other) produce dynamic measuring values which cannot be used for compaction assessment, even if the surface is sufficiently level.

6.2.5 *Resting time of the compacted layer*

The resting time between completion of compaction and measuring pass should be kept as short as possible, because this resting time may cause changes in the dynamic measuring values. The reasons could be site condition (site traffic), weather influences or subsequent settling. Weather and larger resting times between the completion of compaction and the subsequent measuring pass(es) influence the measured values. The longer this period, the more problematic is the establishment of a reference to the original result.

6.3 Roller

6.3.1 *Amplitude*

The theoretical amplitude of the drum is a roller parameter resulting from drum mass, eccentric mass and its eccentricity. The magnitude of the amplitude influences the compaction effect, the measuring depth, the motion behaviour of the drum (operating condition), and consequently the level and range of the dynamic measuring values.

If operating with high amplitude, the compaction effect and measuring depth are higher, the risk of grain crushing and re-loosening of soil near the surface increases, and the drum has a higher tendency to (double) jump than during operation with low amplitude.

6.3.2 *Frequency*

The vibration frequency is to be kept constant during the measuring pass. Frequency fluctuations influence the course of the dynamic measuring values and are not permitted during measuring passes (tolerance range +/− 2 Hz).

6.3.3 *Static line load of drum*

The static line load is a roller parameter which results from the load of the drum and the effective frame weight related to the drum width. The higher the line load, the higher the compaction effect and measuring depth. Deeper areas, therefore, gain in influence on the dynamic measuring values.

Furthermore, the relation between effective frame weight and drum weight influences the motion behaviour of the drum (operating condition), whereby rollers with light frame have a higher tendency to (double) jump.

6.3.4 *Travel speed*

The rolling speed should be between 2 and 6 km/h and must be kept at a constant level throughout the measuring pass. Speed fluctuations influence the course of the dynamic measuring values and are not permitted during measuring passes (tolerance range +/− 0.2 km/h).

6.3.5 *Driving direction forward or backwards*

Measuring passes should only be performed in one direction.

If construction purposes require measuring passes to be performed in both travel directions, hence to and fro comparative passes must be performed beforehand to check whether the level of measuring values obtained during backwards passes differ from the values obtained during forward passes.

6.4 Position and geometry of the rolling track

6.4.1 *Inclinations*

Roller-integrated continuous compaction control can be used on any sloped surface if the measuring roller can be operated there without problems.

Transverse to the dip line of a slope
Inclinations up to 5% have no relevant influence on the measured values. If the inclination becomes steeper, it becomes increasingly difficult to stay exactly in the rolling line.

Parallel to the dip line of a slope
Dynamic measurements at slope gradients up to about 5% lead to almost identical results, for uphill and downhill driving. The steeper the gradient, the greater is the difference between measuring values for uphill and downhill travel.

Therefore, the rolling pattern described in item 8.3 should be applied for steeper gradients.

6.4.2 *Narrow curves*

Measuring passes around narrow curves should be avoided.

6.4.3 *Edge, zones of embankments, excavations*

Since the transducer unit is mounted on one side of the roller, significantly different measuring values may be achieved if – depending on the driving direction – the transducer unit is positioned near the edge of an embankment slope or on the other site, i.e. closer to the middle of the road etc. In general, lower values occur close to the edge of the embankment than farther away from it.

Furthermore, the dynamic measuring values are influenced by:

- height and gradient of the slope
- compaction condition of the embankment

6.5 Measuring depth

The measuring depth depends particularly on the static line load of the drum, the amplitude and frequency, and on the soil.

The following values are achieved, e.g. for the compaction of gravel placed in layers:

- 2 t rollers approx. from 0.4 to 0.6 m
- 10 t rollers approx. from 0.6 to 1.0 m
- 12 t rollers approx. from 0.8 to 1.5 m

It is pointed out that the measuring depth is generally deeper than the compaction depth.

7 CALIBRATION BY COMPACTION ON A TEST FIELD

7.1 Fundamentals

The calibration in connection with a test compaction represents some sort of "approval test" for the soil material used and the roller, and should, with respect of the compaction process, lead to a procedure that provides an optimised result concerning quality, organisation and economic aspects likewise.

The calibration is used for correlating the dynamic measuring values to the required acceptance values obtained by conventional tests, as demanded in other contractual provisions. The calibration is therefore a part of the acceptance test and must be performed by that institute which controls on behalf of the client.

The compaction and measuring passes required for the calibration must, however, be performed by the contractor with the measuring roller as ordered by the site supervisor or client. Measuring roller, measuring system and trained roller driver must therefore be provided by the contractor.

7.2 Comparative tests

Roller-soil-measuring systems form a characteristic unit with the specific amplitude, frequency and travel speed. This unit has to be calibrated with the results from conventional tests.

On principle, it is possible to assign the dynamic measuring values to all results of conventional tests concerning compaction and load-bearing capacity.

The correlation between the dynamic measuring value and the E_{v1} modulus of the static load plate test (with d = 30 cm plate) should be preferred, because it is generally better than the correlation to E_{v2} values. The ratio E_{v2}/E_{v1} is completely unsuitable for a correlation. The correlation between density measurements and dynamic measuring values is also problematic; appropriate experience is required for this. On the other hand, correlation between roller-integrated CCC-values and the modulus $E_{v,dyn}$ from dynamic load plate tests (with the Light Falling Weight Device) are rather promising.

For the reasons stated above, the common calibration procedure is based on the correlation between dynamic measuring values and E_{v1} values (of the static 30 cm load plate). If, however, a correlation is set up with other characteristic values (e.g. with the E_{vd} value of the dynamic load plate), the procedure described is also suitable, but the comparative tests have to be carried out four times.

In order to avoid unacceptable deviations of the correlation, utmost care is required when performing load plate tests. This refers especially to

- evenness
- wet or loose/soft surface
- same measuring points (roller, load plate).

It should also be noted that roller-integrated CCC and static load plate tests involve different measuring depths.

7.3 Selection of a test field

The calibration must be performed on the specific construction section and not in an arbitrary grid farther away.

Commonly, a characteristic section with a length of 100 m and the entire width of the road (or embankment) should be selected as test field within the construction section. This section must correspond to the average properties of the layers to be compacted within the whole construction section. In the case of roads with separated directional tracks, half the crown width is sufficient as width of the test field.

The test field has to be divided in n (integral) roller tracks whereby the maximum track width corresponds to the drum width. The minimum roller track width is determined by the fact that the overlapping should not exceed 10% of the drum width.

7.4 Choice of the measuring roller

The equipment has to be selected by experience considering the layers (multi-layer systems), the material to be compacted (see item 6.2) and the position and geometry of the roller tracks (see item 6.4). The roller parameters as total weight, static line load of the drum, frequencies, amplitudes(s) (see item 6.3) and the roller's ability to compact sloped surfaces have to be particularly considered. Another decision criterion is the desired measuring depth of the roller (see item 6.5).

If the operating condition (double) jump occurs the calibration and the subsequent assessment of the dynamic measuring values become more difficult. Consequently, this possibility should also be considered when choosing the measuring roller (see item 6.1, item 6.3 and appendix).

7.5 Calibration procedure

7.5.1 *Documentation of the existing subgrade*

- Rolling procedure

On the first roller track of the (already compacted) test field a measuring pass has to be carried out in forward direction, followed by a static pass in reverse on the same track. This procedure has to be carried out twice for each roller track.

If the result on the track differs widely from the average of the other roller tracks, further passes must be performed to check whether the compaction along that track can still be increased. After the third additional pass the test must be stopped under all circumstances. This test result is the basis for further measures to be applied (e.g. investigation of the existing soil).

If jump operation occurs during a measuring pass, the already selected roller parameters have to be maintained, and the documentation of the layer quality has to be continued.

- Measurement with the load plate

In the test field the E_{v1} value has to be measured immediately after the measuring pass of the roller. This has to be performed at locations with low, medium and high dynamic measuring values.

If unavoidable (double) jump operation occurred to a high degree during the measuring passes the E_{v1} value has to be determined also in these areas.

7.5.2 *Test compaction*

By means of roller-integrated continuous compaction control the optimum thickness of the layers, an optimised compaction procedure, and the relevant type and number of roller passes for final compaction have to be determined.

During this process different amplitudes and static passes should be considered; amplitude, frequency and travel speed have to be fixed then for the measuring passes.

In the test field the material to be compacted has to be placed and distributed in the chosen layer thickness.

- Rolling procedure

On the first track a measuring pass, which is generally a compaction pass has to be performed in forward direction, followed by a static reverse pass on the same track. This procedure has to be repeated, until no noticeable increase of the dynamic measuring values is evident. Furthermore, this testing step has to be repeated for each track after transferring the measuring roller from one track to another.

If significant jump operation occurs during a measuring pass, the described testing sequence has to be interrupted and the measuring pass has to be repeated with lower amplitude and/or a higher travel speed. If (double) jump of the drum cannot be prevented even after these measures, an additional calibration has to be performed for this operating condition, or a more suitable roller (see item 6.3.3) should be used.

It is, however, absolutely necessary that the roller parameters finally chosen are kept constant for all further measuring passes. Tracks measured with other roller parameters must be measured at least once with the roller parameters chosen last.

Within the following three days an additional measuring pass has to be carried out for each track.

- Measurement with the load plate

Immediately after the test passes, which were carried out with the last chosen roller parameters, a total of nine measurements of the E_{v1} value have to be conducted in the test field at locations exhibiting low, medium and high dynamic measuring values, where no (double) jump operation occurred. It is up to the judgment of the tester after which passes the measurements will be performed. The measuring points have to be chosen such that a rather wide range of dynamic measuring values is obtained. It is permitted to perform some of the measurements after those additional measuring passes which have to be performed within the following three days.

If jump operation during the measuring passes was unavoidable to a high degree, six additional measurements of the E_{v1} value have to be carried out in these zones,

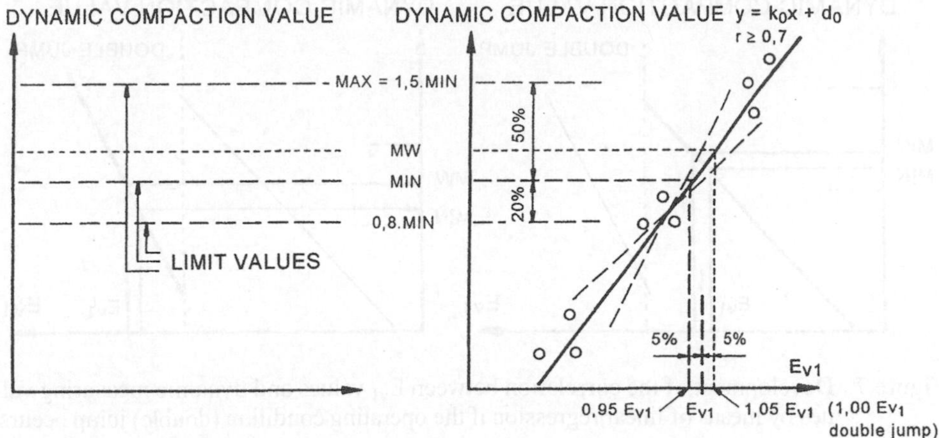

Figure 6. Development of the correlation between E_{v1} values and dynamic measuring values by means of linear regression.

namely two tests on each location showing low, medium and high dynamic measuring values.

7.6 Development of correlations

The correlation has to be determined as a linear regression (mean straight line) according to the formulas in the appendix (item 1) and be drawn in a diagram presented (Figure 6). The correlation coefficient must be $\geqslant 0.7$; otherwise additional measurements are required.

The minimum value (MIN) has to be determined at 95%, the mean value (MW) at 105% (during jump operation at 100%) of the required E_{v1} value. Moreover, the value at 80% of the minimum value (0.8 MIN) and the maximum value (MAX) at 150% of the minimum value have to be specified too – Figure 6.

If the development of a correlation is absolutely necessary also during (double) jump operation (see item 7.5.2) a specific regression function must be determined and incorporated in the diagram. The limit or the limit range in which the (double) jump operation occurs must be specified. In this case, the straight regression line for that range in which (under the specific operation condition) the required E_{v1} value is lying must be used for the correlation – Figure 7.

The determined straight correlation line(s) is (are) only valid for the investigated characteristic unit (unit roller – soil – measuring system with chosen amplitude, frequency and travel speed).

7.7 Documentation

A thorough documentation is required and must at least include the details referred to in item 11.3.

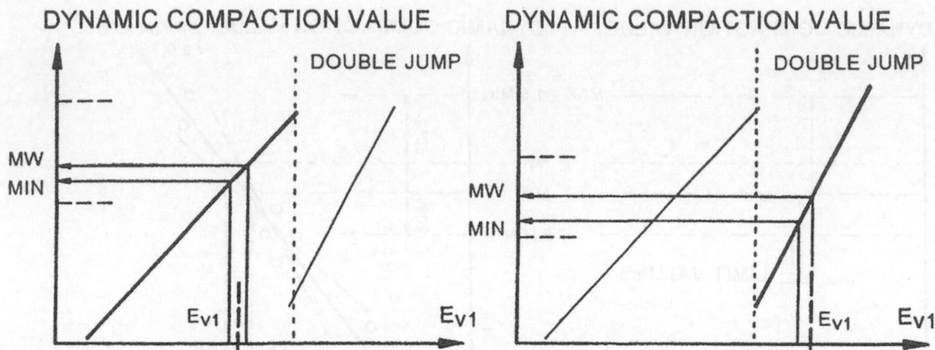

Figure 7. Development of the correlation between E_{v1} values and dynamic measuring values by means of linear regression if the operating condition (double) jump occurs.

8 CONDUCTION OF ROLLER-INTEGRATED CONTINUOUS COMPACTION CONTROL

8.1 Fundamentals

The application of roller-integrated CCC makes a continuous and work-integrated control and acceptance testing possible that covers the entire surface. If CCC is used for control and acceptance testing the procedure described below has to be followed.

8.2 Subdivision in measuring fields

The whole construction section has to be divided into reasonable measuring fields covering the entire area. Intersections are not permitted.

The determination of the measuring fields for the acceptance test has to be made jointly by the contractor and the controller (on behalf of the client).

The size of the measuring fields as well as the lengths and widths of the tracks should correspond to those of the test field.

8.3 Rolling pattern

Measuring passes (i.e. dynamic passes) have to be performed in forward direction on one track of the test field. Reverse travel should be performed statically (see also item 6.3.5). If the measuring procedure is completed on one track, the roller must be transferred to the next track.

8.4 Compaction passes and measuring passes

Compaction passes are used for compaction and have generally been performed in accordance with the compaction procedure specified in item 7 for test compaction.

Measuring passes are roller passes with the measuring roller during which the dynamic measuring values are documented. In this connection it is absolutely necessary that the roller parameters fixed according to item 7 are kept constant during all measuring passes.

In order to guarantee a work-integrated compaction control, measuring passes should be at the same time compaction passes required for the construction task.

8.5 Documentation

For documentation the items 11.3.1 and 11.3.3 apply accordingly, whereby only the area plot as required in item 11.3.3 must be printed out on paper. The track plots of the individual measuring passes must be secured in a suitable storage medium and presented as paper printout only on demand (in excerpts).

9 TESTS

9.1 Fundamentals

If this RVS is incorporated in the construction contract, no limit value, mean value etc. of a (commercial) measuring system should be specified or required in the contract. Formulations for contractual provisions and tender texts are recommended in the appendix (item 2).

9.2 Control tests

Control tests are tests (measuring passes) performed by the contractor to determine whether the quality of his work corresponds to the contractual requirements. If the contractor takes the client's tester for assessing the dynamic measuring values, the measuring passes can be utilised for the acceptance, supposed that the values required in item 9.3. are reached. That part of the control test applied for acceptance should then be treated as acceptance test with all consequences.

9.3 Acceptance tests

9.3.1 *General*

Acceptance tests are necessary to check whether the quality of executed work corresponds to the contract. Their results are the basis for acceptance and accounting. The acceptance tests must be ordered by the client, taking into account that the tests should take place in a work-integrated way during construction. The determination of the limit values according to item 7 and the assessment of the dynamic measuring values obtained from those measuring passes, which form the basis of the acceptance test, have to be carried out by authorised, state or affiliated institutes.

The compaction and measuring passes required for the acceptance should, however, be performed by the contractor with the measuring roller by order of the site supervisor/tester. Measuring roller, measuring system and trained roller driver have to be provided by the contractor.

Jump operation is to be treated separately from other operating conditions (contact and lift-off).

Systematic deviations have to be treated separately. Examples for systematic deviations are given in the appendix (item 1).

The acceptance is valid for all areas in which the values required in item 9.3.2, item 9.3.3 and 9.3.4 are achieved. However, if the measured values are lower than the required ones on more than half of a track or test field, the acceptance test must be repeated for the whole track or the whole test field after having taken appropriate technical measures to reach the required values fixed in the contract.

In case of doubt, the tester/controller may order a so-called "re-calibration" for clearly defined areas upon consultation with the client and he must carry out the corresponding comparative tests. This procedure corresponds to item 7, but with significantly reduced test volume. The results of the re-calibration have to be used for the acceptance then.

9.3.2 *Indirect test of the compaction quality*

(1) The mean measuring value of a measuring pass should not be lower than the mean value (MW) specified in item 7.6.

(2A) If the measured minimum value within a measuring pass is equal to or higher than the specified minimum value (MIN), item 9.3.3 section (3A) has to be applied for judging the uniformity of the compaction.

(2B) Measured dynamic measuring values should not be lower than the specified minimum value (MIN) on a maximum length of 10% of the track during a measuring pass. The measured minimum value should not, however, be lower than 80% of the specified minimum value (0.8 MIN). In the case of such "slightly lower values", item 9.3.3, section (3B) has to be applied for judging the uniformity of the compaction.

If a clear relationship between grain crushing or (re-)loosening and dynamic measuring values is proved, an absolute maximum value of the dynamic measuring values can be also determined/ordered by the tester.

9.3.3 *Uniformity of compaction*

(3A) The standard deviation (relating to the mean value) should not exceed 20% during a measuring pass, as far as item 9.3.2 section (2A) is applicable.

(3B) The maximum value measured during a measuring pass should not exceed the maximum value (MAX) determined according to item 7 at 150% of the specified minimum value as far as item 9.3.2, section (2B) is applicable.

9.3.4 *Increase of compaction degree*

(4) If the mean value of a measuring pass is more than 5% higher than the mean value of the preceding measuring pass, the measuring passes should be continued,

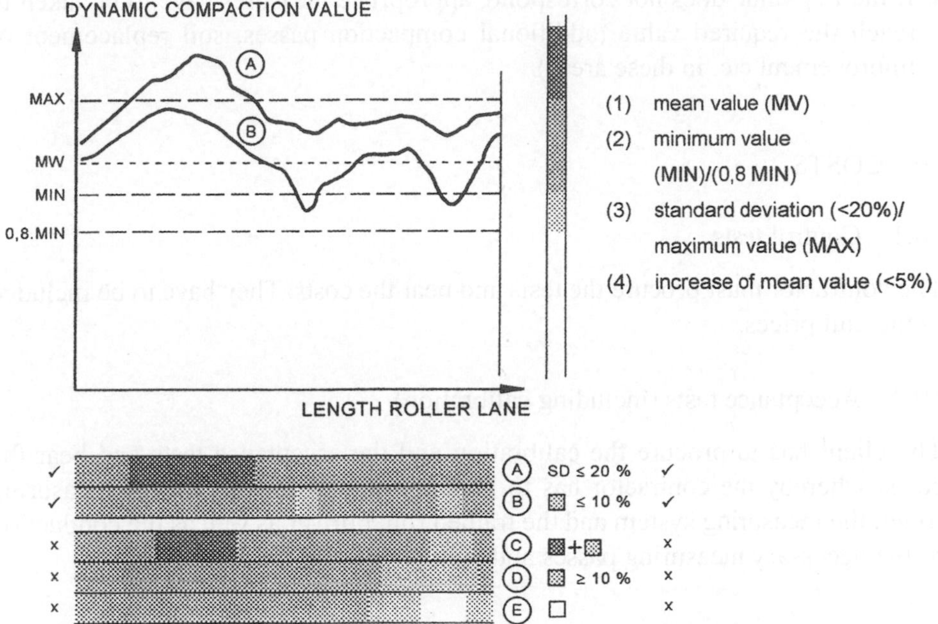

Figure 8. Example: Track plot (courses of measuring values (A), (B)) and area plot (courses of measuring values (A), (B), (C), (D), (E)). Courses of measuring values (A) corresponds to the acceptance criteria according to item (1), item (2a) and item (3A); course of measuring value (B) corresponds to the acceptance criteria according to item (1), item (2B) and item (3B). Courses of measuring values (C), (D) and (E) do not correspond to all acceptance criteria.

until the increase of the mean value (compared with the preceding measuring pass) is only max. 5%.

If the mean value drops during two subsequent measuring passes, the tester may order that further roller passes have to be stopped. A clear connection between the drop of the dynamic measuring values and a re-loosening of the soil or crushing of grains must, however, be evident.

9.3.5 *Acceptance test on small construction sites*

On small construction sites and in places where, for economic and/or technical reasons (confined space, construction work under traffic etc.) the calibration cannot be reasonably performed according to item 7, the following procedure is applied:

- Compaction with the measuring roller is continued until no compaction increase according to item 9.3.4 is registered.
- The E_{v1} value is measured at the weakest point(s).
- If this E_{v1} value corresponds to the required value, compaction of the next layer can start. For documenting the acceptance test, item 8.5 applies accordingly.

- If the E_{v1} value does not correspond, appropriate measures have to be taken to reach the required value (additional compaction passes, soil replacement or improvement etc. in these areas).

10 COSTS

10.1 Control tests

The contractor must procure the tests and bear the costs. They have to be included in the unit prices.

10.2 Acceptance tests (including calibration)

The client has to procure the calibration and the acceptance tests and bear the costs, whereby the contractor has to include the cost for driving the measuring roller, the measuring system and the trained roller driver as well as the conduction of the necessary measuring passes and include these in his unit prices.

11 APPENDIX

11.1 Explanatory notes

Ad item 5.4: Review of the reproducibility of series of measurements

The reproducibility of series of measurements obtained during subsequent measuring passes has to be checked on a section of approx. 100 m length which shows rather non-uniform subsoil properties.

During this process a measuring pass has to be performed in this particular section in forward drive, followed by a static pass in reverse on the same track. After another measuring pass in forward direction on the same track the roller has to be turned at the end of the track, and another measuring pass has to be performed on the same track in forward motion, followed by a static pass in reverse on the same track. The control procedure has to be completed with another measuring pass on this track in forward direction.

Reproducibility is given if all four series of measurements indicate a uniform characteristic course of measuring values and the level of the measuring values varies only slightly, i.e. increases only due to a further compaction or decreases due to surface-near re-loosening or grain crushing.

Ad item 6.1: Operating conditions of the drum

Depending on soil stiffness, and on different roller parameters and travel speed, the excited drum of a vibratory roller shows different behaviour that can be described by five operating conditions. Roller-integrated CCC has to be considered in connection with these operating conditions, since the motion behaviour of the roller drum has a significant influence on the dynamic measuring values. The operating conditions are characterised by a possible contact loss of the drum and the period duration of the drum motion. Terms and definitions are given in Table 2.

The operating condition "*continuous contact*" is rare and occurs only in the case of very low soil stiffness. The dynamic measuring value is also low. Conventional rollers are generally designed in such a manner that their drums are in the operating condition "*partial uplift*" at common soil stiffness. The dynamic measuring value rises with increasing soil stiffness. The transition of the measuring value from contact to lift-off occurs continuously so that these two operating conditions can be combined, thus representing the usual type of operation.

The operation condition "*(double) jump*" occurs at high soil stiffness and has a significant influence on the behaviour of the dynamic measuring values. The transition between partial uplift and (double) jump is characterised by the fact that the measuring values decrease in general and change to a lower level. However, with increasing stiffness, the dynamic measuring value increases again under this operating condition. Therefore, it is possible and reasonable to use roller-integrated CCC also during jump operation. However, measuring values being determined during the so-called (double) jump operation have to be automatically identified and noted as such by the measuring system (with registered location). (double) jump operation is also identifiable by the increased shaking effect and acoustically by a deep doom.

The operating conditions "*rocking motion*" (in this process the drum performs a swaying motion, i.e. it bounces on the ground alternatingly on the right and on the

Table 2. Definition of the operation conditions of vibratory rollers and qualitative relationship with the soil stiffness.

drum motion	Interaction drum-soil	operating condition	soil contact force	application of CCC	soil stiffness	roller speed	drum amplitude
periodic	continuous contact	CONT. CONTACT		yes	low	fast	small
	periodic loss of contact	PARTIAL UPLIFT		yes	↓	↑	↓
		DOUBLE JUMP		yes			
		ROCKING MOTION	left right	no			
chaotic	non-periodic loss of contact	CHAOTIC MOTION		no	high	slow	large

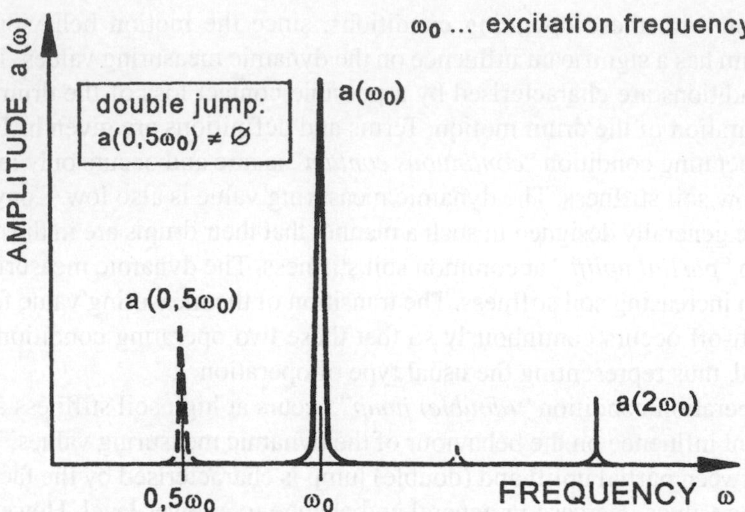

Figure 9. Definition of jump operation.

left side) and "*chaotic motion*" only occur at extremely high soil stiffness in combination with unfavourable roller parameters and low travel speed. It should automatically be ruled out because either no reasonable measuring values or no dynamic measuring values at all can be obtained. These two types of operating condition can be identified by the fact that the roller is no more controllable, i.e. it is absolutely impossible to stay within the track.

Ad item 7.6: Linear regression

In general, the following calculation and graphical processing is integrated in the software of the specific measuring system.

On the one hand the dynamic measuring values are assumed in dependence on the comparable conventional values, and on the other hand the comparable values are considered dependents on the dynamic measuring values. This results in two different straight regression lines intersecting at the centre of gravity of the point cloud (corresponds to the coordinates with the mean values $\bar{x}$ and $\bar{y}$ of the random samples).

Furthermore, another straight line, the so-called "straight centre line" is needed. It also runs through the centre of gravity, and its gradient is derived from the angle between the straight regression lines (bisecting line of the angle).

On the abscissa (x) the comparative values (E_{v1}) and on the ordinate (y) the dynamic measuring values have to be plotted.

The two straight regression lines are defined as follows:

$y = K_1\,x + d_1 \ldots$ dynamic measuring values depending on comparative values

$y = K_2\,x + d_2 \ldots$ comparative values depending on dynamic measuring values

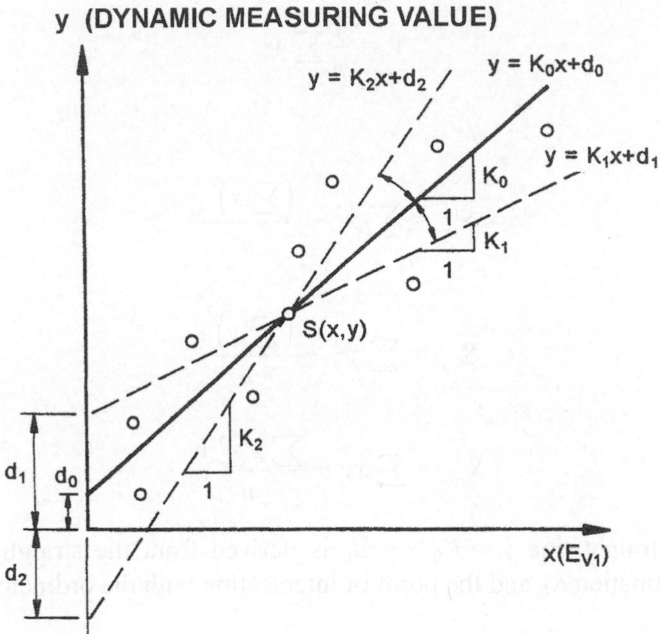

Figure 10. Linear regression analysis.

whereby K_0 represents the inclination of the regression line and d_0 the point of intersection with the ordinate.

$$K_1 = \frac{S_{xy}}{S_{xx}} \tag{1}$$

$$K_2 = \frac{S_{yy}}{S_{xy}} \tag{2}$$

$$d_1 = \bar{y} - K_1\bar{x} \tag{3}$$

$$d_2 = \bar{y} - K_2\bar{x} \tag{4}$$

$\bar{x}$ and $\bar{y}$ represent the mean values of the random samples x and y; n is the number of the pair of random samples:

$$\bar{x} = \frac{1}{n}\sum x \tag{5}$$

$$\bar{y} = \frac{1}{n}\sum y \tag{6}$$

Furthermore:

$$S_{xx} = \sum x^2 - \frac{\left(\sum x\right)^2}{n} \tag{7}$$

$$S_{yy} = \sum y^2 - \frac{\left(\sum y\right)^2}{n} \tag{8}$$

$$S_{xy} = \sum xy - \frac{\sum x \sum y}{n} \tag{9}$$

The centre straight line $y = K_0\, x + d_0$ is derived from the straight regression lines, the inclination K_0 and the point of intersection with the ordinate d_0 result as follows:

$$K_0 = \tan\left(\frac{\arctan K_1 + \arctan K_2}{2}\right) \tag{10}$$

$$d_0 = \bar{y} - K_0\bar{x} \tag{11}$$

The correlation coefficient r is equal for the two straight regression lines and the centre straight line:

$$r = \frac{S_{xy}}{\sqrt{S_{xx}\, S_{yy}}} \tag{12}$$

Ad item 9.3.1: Examples for systematic deviations of measuring values (Figure 11).

11.2 Contractual terms and recommendations for tender texts for CCC application

If roller-integrated continuous compaction control (CCC) is used for control and acceptance testing, the following contractual terms are recommended to be included in the specific construction contracts and the tender:

Checking the compaction and/or load bearing capacity (control and acceptance test) is to be carried out using the roller-integrated continuous compaction control (CCC) following the procedure according to these technical provisions.

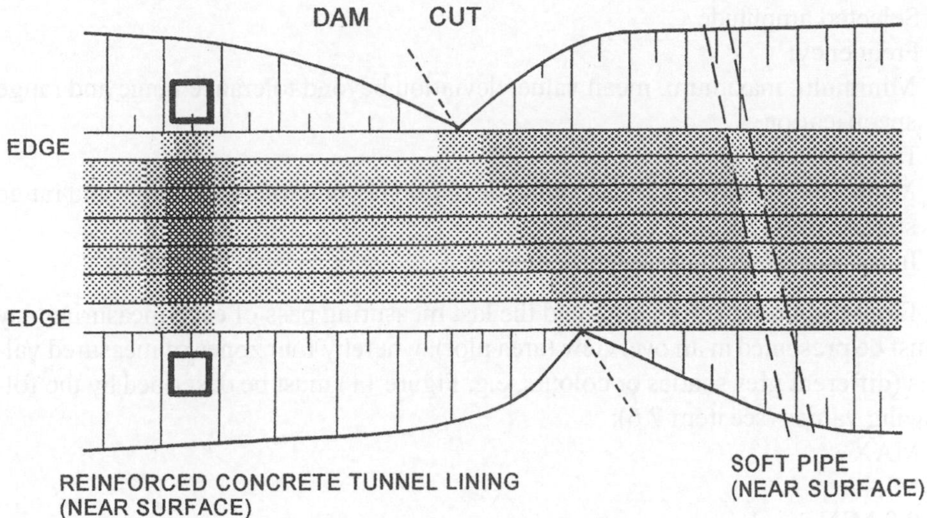

Figure 11. Area plot. Systematic deviations: Transition zone between embankment and cut, edge tracks, (soft) pipeline and (stiff) culvert.

Measuring roller, measuring system and a trained roller driver (i.e. familiar with CCC) have to be provided by the contractor. Prerequisites and requirements have to be observed according to these technical provisions, item 5. Measuring passes which are used as basis for the acceptance (also measuring passes for calibration) have to be carried out under instruction of the tester/site supervisor ordered by the client in accordance with these technical provisions, item 7, item 8 and item 9.

11.3 Documentation

11.3.1 *General*

Generally required details should be listed in a test form. They have to be documented and stored for each test field.

11.3.2 *Calibration procedure in connection with the test compaction*

- Rolling pattern
- Sequence and number of compaction and measuring passes per track
- Change of amplitude and/or travel speed of the roller (with reasons)
- Comparative tests: Locations, allocation to the specific measuring pass

11.3.3 *Measuring passes*

Prior to each measuring pass the following must be automatically documented by the measuring system:

- Course of the dynamic measuring values (track plot)
- Minimum, maximum, mean value, standard deviation in percent (referring to the mean value) of the dynamic measuring values

- Selected amplitude
- Frequency:
  Minimum, maximum, mean value, deviation beyond tolerance range and range specification
- Travel speed:
  Minimum, maximum, mean value, deviation beyond tolerance range and range specification
- Jump operation: range specification

From each track of the test field the last measuring pass of each measuring day must be presented in an overview (area plot), whereby four zones of measured values (different grey shades or colours, e.g. Figure 11) must be described by the following values (see item 7.6):

- MAX
- MIN
- 0.8 MIN

REFERENCES

Adam, D.: *Roller-integrated continuous compaction control of soils by means of vibratory rollers*. (in German). Doctor's Thesis. Technical University of Vienna, 1996.

Brandl, H., and Adam, D.: *Roller-integrated continuous compaction control (CCC) with vibratory rollers* (in German). Volume 506 of the series "Road Research" published by the Federal Ministry for Transportation, Innovation, and Technology, Vienna 2001.

Brandl, H. and Adam, D.: *Basics and Application of the Dynamic Load Plate Test in Form of the Light Falling Weight Device*. A.W. Skempton Memorial Conference. Proc. of Imperial College, London, 2004.

Kopf, F.: *Roller-integrated continuous compaction control of soils by means of different dynamic rollers*. (in German). Doctor's Thesis. Technical University of Vienna, 1999.

RVS 8S.02.6: Technical contractual provisions: Earth works. Roller-integrated continuous compaction control (CCC). Vienna, 2000.

10

Crushed Concrete Aggregate as a Backfill Material for Civil Engineering Soil Structures

F. Tatsuoka &Y. Tomita
Tokyo University of Science, Chiba, Japan

L. Lovati
Politecnico di Torino, Torino, Italia

U. Aqil
National Institute of Rural Engineering, Ibaragi, Japan

ABSTRACT: Drained triaxial compression tests were performed on crushed concrete aggregate to investigate the feasibility of its use as a backfill material for civil engineering soil structures requiring a high stability while allowing a limited amount of deformation, such as embankments and soil retaining walls supporting highway and railway. It is shown that the compressive strength, q_{max}, when well compacted is similar to, or even higher than, that of typical selected highest-class backfill material, such as natural well-graded gravelly soil. The effect of confining pressure on q_{max} is similar as natural well-graded gravelly soil. q_{max} is essentially a unique function of compacted dry density, ρ_d, irrespective of water content at compaction, w. The pre-peak stiffness, E_{50}, is also a function of ρ_d, but the relation is slightly affected by w. The values of q_{max} and E_{50} are very sensitive to the degree of compaction, noticeably more than natural well-graded gravelly soil. It is indicated that, when well compacted, crushed concrete aggregate can be used as a high-class backfill material.

1 INTRODUCTION

In many developed countries, including Japan, a great amount of crushed concrete aggregate is being and will be produced, mostly from demolished steel-reinforced concrete structures. The use of crushed concrete aggregate as the backfill material of soil structures, such as embankments and soil retaining walls, replacing selected high-class backfill soil (i.e., natural well-graded gravelly soil) is being highly required for the following reasons:

1) Currently, most crushed concrete aggregate is used as the road base material. However, it is estimated that such a secondary use as above will soon become insufficient.

2) Dumping of concrete scrap in remote places is becoming unacceptable due to its too high environmental impact.
3) An intolerable level of energy is needed to crush and treat concrete scrap to produce recycled fine and coarse aggregates with which concrete can have essentially the same strength as the one produced by using fresh natural aggregates.

It is widely considered that crushed concrete aggregate is a kind of inferior construction material having strength and stiffness much lower than those of selected natural backfill soils while exhibiting large residual strains during a long life time. Indeed, it is the case when not well compacted (e.g., Mizukami et al., 1998). However, Aqil et al. (2005a, b) showed that, when well compacted, crushed concrete aggregate exhibits compressive strength equivalent to that of highest-class natural backfill material. The present study is a follow-up of that study, performed to evaluate effects of the degree of compaction and compaction energy level on the strength and deformation characteristics of crushed concrete aggregate compacted at different water contents. The effects of compaction were compared between crushed concrete aggregate and a typical natural well-graded gravely soil having the same grading curve to highlight the effects of the specific particle constitution of crushed concrete aggregate (i.e., a thin layer of relatively soft and weak mortar covering stiff and strong coarse core particles).

2 TEST METHOD AND MATERIALS

2.1 Triaxial apparatus

An automated triaxial testing apparatus (Fig. 1) was used. The specimen was 10 cm in diameter and 20 cm high. The deviator load was measured with an internal load cell. Local axial and lateral strains reported in this paper are those obtained by averaging the outputs from, respectively, a pair of Local Deformation Transducer (LDTs) (Goto et al., 1991) and three clip gauges set at 5/6, 1/2 and 1/6 of the specimen height from the bottom. The volumetric strains were obtained from local axial and lateral strains. External axial strains were also obtained by measuring axial displacements of the loading piston with a displacement transducer (LVDT) set outside the triaxial cell to evaluate axial strains towards values exceeding the measuring limit of the LDTs.

2.2 Test materials

Two types of crushed concrete aggregate were used (see Fig. 2 and Table 1). The first one (recycled railway embankment aggregate, RREA) was retrieved from the subbase of a geogrid-reinforced embankment constructed as a temporary railway structure, which was demolished in July 2002. The other one (recycled electricity pole aggregate, REPA) was obtained by fresh-crushing the concrete of electricity poles (Aqil et al., 2005b). Their grading curves were made similar to eliminate the effect of gradation characteristics on the test results.

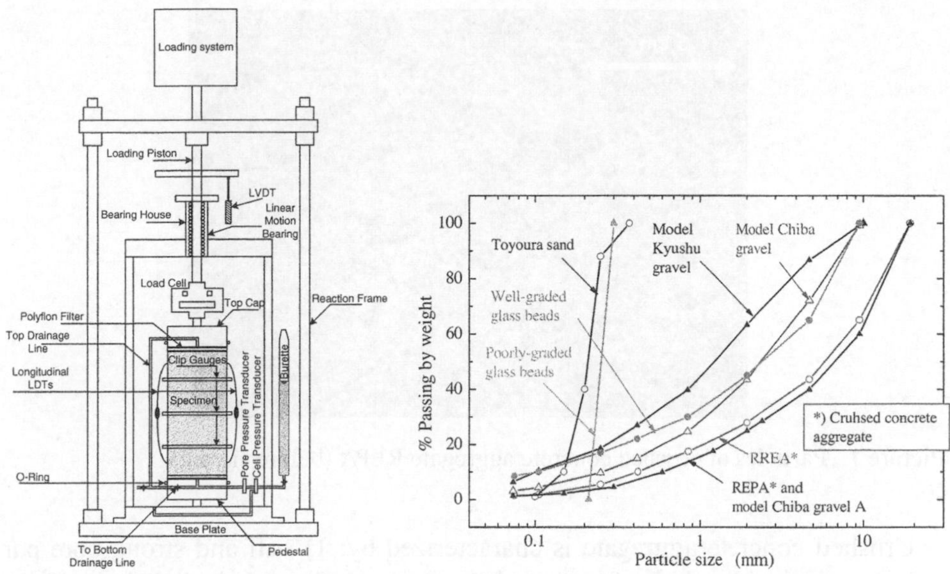

Figure 1. (left) Triaxial testing apparatus used in the present study (Aqil et al., 2005b).
Figure 2. (right) Grain size distribution curves (before compaction) of tested materials.

Table 1. Materials used in the present study.

Geomaterial name	Particle constitution	Particle shape	G_s	D_{max} (mm)	D_{50} (mm)	U_c	Fines content, F_C(%)
RREA[1),2),4)]	A think soft & weak layer covering stiff & strong cores	Angular	2.65	19	5.84	18.76	1.32
REPA[3)]			2.45	19	6.67	15.83	–
Model Chiba gravel A[3)]	Stiff and strong particles	Angular	2.74				
Model Chiba gravel[1)]			2.74	9.50	2.46	13.2	3.1
Model Kyushu gravel[1)]		Angular	3.11	9.50	1.22	18	6.4
Toyoura sand[1)]		Angular	2.65	0.375	0.209	1.46	–
Poorly graded glass beads[1)]		Sphere	2.50	0.30	0.253	1.18	–
Well graded glass beads[1)]			2.50	4.80	2.48	36.57	8.50

1), 2), 3) & 4): materials used in test series 1, 2, 3 & 4, respectively.

Picture 1. Particles of crushed concrete aggregate REPA (before sieving).

Crushed concrete aggregate is characterized by: 1) stiff and strong core particles covered with a thin soft and weak mortar layer; 2) a relatively large uniformity coefficient; and 3) relatively angular particle shapes (Photo. 1). To evaluate the effects of these factors on the strength and deformation characteristics of crushed concrete aggregate, a set of consolidated drained triaxial compression (CD TC) tests were performed on several typical granular materials often used in practice and research in addition to RREA and REPA (see Fig. 2 and Table 1); i.e., a poorly-graded fine quartz-rich sand, Toyoura sand, and poorly- and well-graded glass beads as well as three natural well-graded gravelly soils (model Chiba gravel, model Chiba gravel A and model Kyushu gravel). These gravelly soils were obtained by sieving out particles larger than 10 mm or 19 mm from typical natural gravelly soils from quarries (i.e., Chiba gravel, consisting of crushed sandstone; and Kyushu gravel, consisting of crushed gabbro), which are categorized as the highest-class backfill material for civil engineering soil structures. Model Chiba gravel A has the same grading curve as REPA.

Figure 3 shows the compaction curves of RREA and REPA obtained by using modified Proctor compaction energy level ($E1 = 2.48\,\text{Nm/cm}^3$). The maximum dry densities $\rho_{d.max}$ of these materials are similar to those of many other crushed concrete aggregates produced in Japan when compared for the same water absorption ratio, Q, evaluated by the Japanese Industrial Standard A 1110-1999 (Fig. 4). It is also the case with. As seen from Fig. 4, the crushed concrete aggregates have generally much smaller $\rho_{d.max}$ values than natural well-graded gravelly soil (consisting of stiff and strong particles). This is due partly to the inclusion of mortar (cement paste plus fine aggregate, having a lower specific gravity, G_s) and largely to a higher compacted void ratio (as explained later). Despite a lower G_s value, the $\rho_{d.max}$ value of REPA is much higher than that of RREA (Fig. 3). This may be due likely to a higher quality of the source concrete of REPA than RREA, which have

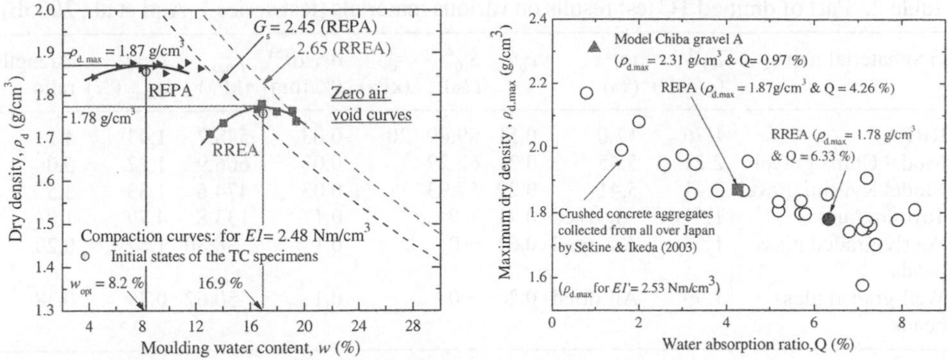

Figure 3. (left) Compaction curves of RREA and REPA (revised from Aqil et al., 2005b).
Figure 4. (right) $\rho_{d.max}$ – water absorption ratio relations of RREA, REPA and model Chiba gravel A and other crushed concrete aggregates reported by Sekine and Ikeda (2003).

resulted in a more amount of mortar that survived the crushing process (resulting into a lower G_s value) and then a more amount of mortar was crushed during the compaction process (resulting into a higher $\rho_{d.max}$ value).

2.3 Test programme

Results from the following four test series are reported in this paper:

1) *series 1* to evaluate effects of particle property in terms of grading characteristics, particle shape and particle constitution (Aqil et al., 2005a, b);
2) *series 2* to evaluate effects of confining pressure (Aqil et al., 2005a, b);
3) *series 3* to evaluate effects of compacted density and moulding water content; and
4) *series 4* to evaluate residual strains by sustained and cyclic loading (Aqil et al., 2005a, b).

All the specimens, moist as compacted, were subjected to automatic isotropic compression at a pressure rate of 0.5 kPa/min by using a computer-aided feedback system, followed by drained TC loading at a fixed axial strain rate.

The specimens of RREA and model Chiba gravel for test series 1, 2 and 4 and model Kyushu gravel for series 1 were prepared by manual tamping in five even sub-layers in a cylindrical split mould (10 cm in inner-diameter and 20 cm in inner height) using $E1$=2.48 Nm/cm³ at w around the respective w_{opt} (Table 2). The specimen taken out from the mould was self-supporting due to suction. Side drainage of vertical filter paper strips was arranged around the specimen to ensure a uniform distribution of partial vacuum applied as effective confining pressure. The specimens of REPA and model Chiba gravel A for test series 3 were prepared in the same way as above but by using three different compaction energy levels (explained later). Toyoura sand (relative density, D_r = 70%), poorly-graded glass beads (D_r = 70%)

Table 2. Part of drained TC test results on various materials (test series 1, Aqil et al., 2005b).

Geomaterial name	ρ_d [1] (g/cm^3)	w [2] (%)	e_o [3]	S_{r0} [4] (%)	σ'_c (kPa)	$d\varepsilon_v/dt$ [5] (%/min)	q_{max} (kPa)	ε_v at q_{max} (%)	Strength ratio
RREA	1.76	17.0	0.51	89.09	20	0.03	548.7	1.43	4.1
Model Chiba gravel	2.21	5.45	0.24	62.27		0.03	666.9	1.22	5.0
Model Kyushu gravel	2.41	5.41	0.29	57.93		0.03	474.6	1.33	3.5
Toyoura sand	1.55	Air dried	0.71	~0		0.1	133.8	4.79	1.0
Poorly-graded glass beads	1.54	Air dried	0.62	~0		0.1	38.80	1.87	0.29
Well-graded glass beads	1.99	Air dried	0.26	~0		0.1	50.62	0.39	0.38

1) compacted dry density; 2) moulding water content; 3) initial void ratio; 4) initial degree of saturation; 5) axial strain rate.

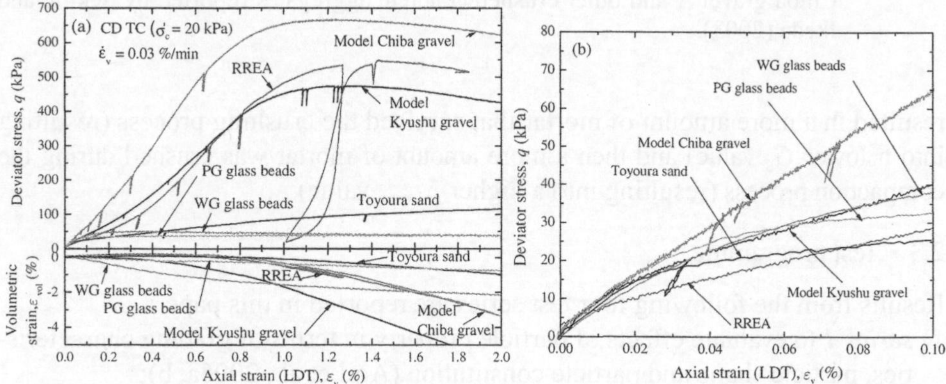

Figure 5. Stress-strain behaviour from test series 1 (Aqil et al., 2005b); a) a large strain range; and b) a small strain range.

and well-graded glass beads ($D_r = 94\%$) for series 1 were prepared by pluviating air-dried particles through air into a split mould (10 cm in inner diameter and 20 cm in inner height) set in the triaxial cell. The top and bottom ends of the specimen were not lubricated but in contact with the rigid flat stainless steel top cap and pedestal. The confining pressure, σ'_c, was equal to 20 kPa in series 1 and 4, 20–90 kPa in series 2 and 30 kPa in series 3.

3 TEST RESULTS

3.1 Effects of particle property

Figure 5a compares the stress-strain relations from test series 1. Both glass beads and Toyoura sand exhibit significantly lower peak strengths than the two types of natural

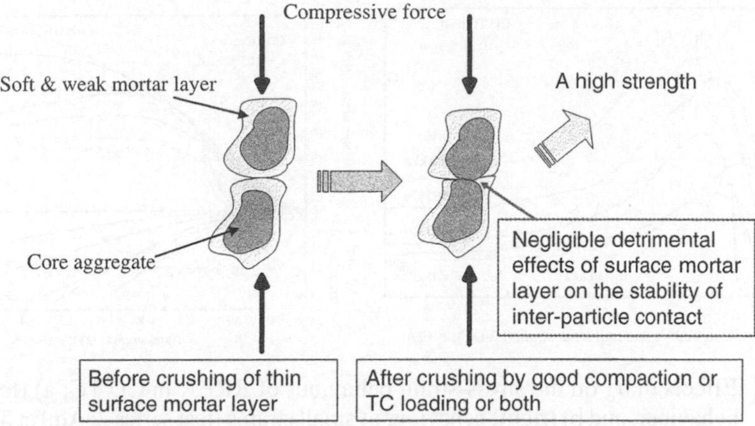

Figure 6. Likely changes by compaction in inter-particle contact conditions of crushed concrete aggregate.

well-graded gravelly soil (consisting of angular stiff and strong particles). These low peak strengths can be attributed to a uniform grading (Toyoura sand), the spherical particle shape (well-graded glass beads) and both (poorly-graded glass beads). On the other hand, as seen from Fig. 5b, the initial stiffness of these weak materials is not particularly low. These two different features can be explained by a very high dependency of the tangent stiffness on the strain level with these materials. The peak strength of RREA is surprisingly high and similar to those of the two highest-class natural backfill materials, model Chiba and model Kyushu gravels. The most distinctive negative feature of crushed concrete aggregate, different from natural aggregates, is that a relatively soft and weak layer of mortar is adhering to the surface of stiff and strong core aggregates (gravel particles) (Fig. 6). It appears that, when subjected to TC loading, some crushing of mortar layer takes place at the inter-particle contact points, resulting in a relatively low initial stiffness (Fig. 5b). Then, a better inter-particle contact condition between adjacent stiff cores of the aggregate develops, resulting in an increase in the tangent stiffness and ultimately a high compressive strength (Fig. 5a). It seems therefore that possible negative effects of thin surface mortar layer on the compressive strength can be effectively alleviated by better compaction.

3.2 Effects of confining pressure

To confirm whether the compressive strength of crushed concrete aggregate is still as high as the highest-class backfill material (i.e., natural well-graded gravelly soil) when the confining pressure, σ'_c, becomes higher than 20 kPa, a set of CD TC tests at higher σ'_c values were performed on RREA specimens densely compacted by energy level $E1$ at $w \approx w_{opt}$. The test results are presented in Fig. 7. Figure 8 shows the compressive strengths q_{max}, the residual strengths q_{res} (defined as q at $\varepsilon_v = 10\%$),

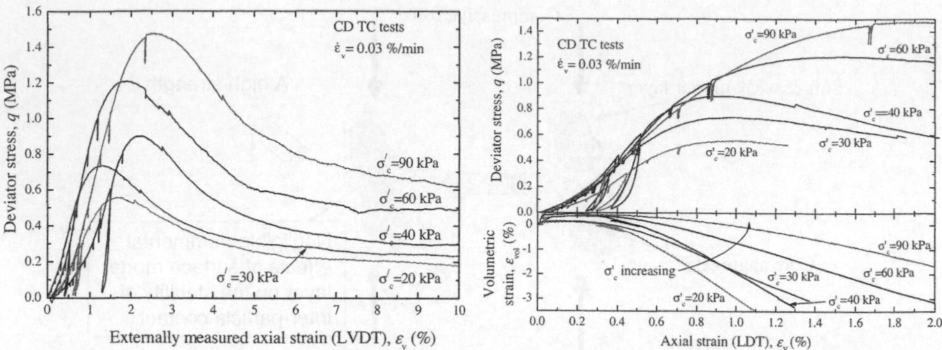

Figure 7. Effects of σ_c' on the stress-strain behaviour of RREA in CD TC; a) (left) overall behaviour; and b) (right) behaviour at small strains (test series 2; Aqil et al., 2005b).

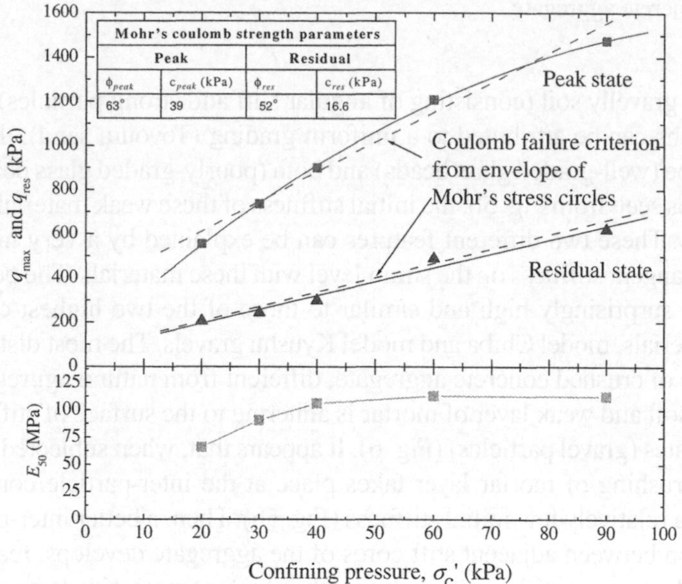

Figure 8. Relationship among peak and residual strengths, E_{50} and confining pressure, RREA (Aqil et al., 2005b).

the E_{50} values plotted against the respective σ_c' value. The solid curves presented in Fig. 8 are those obtained by directly fitting to the data points. It may be seen that both q_{max} and q_{res} increase with an increase in σ_c', while q_{max} increases in a more non-linear fashion than q_{res}, all in a similar way as natural well-graded gravelly soil. The angle of internal friction, ϕ_{peak}, and the cohesion intersect, c_{peak}, as well as the residual friction angle, ϕ_{res}, and the residual cohesion intersect, c_{res} (listed in Fig. 8) were obtained by fitting the linear Coulomb failure criterion to the respective set of Mohr's circles at the peak and residual states. The straight broken lines presented in Fig. 8 represent these linear failure envelops. The high values of ϕ_{peak} and c_{peak} indicate that the peak shear strength of RREA is much higher than those of ordinary poorly-graded soils not only at $\sigma_c'=20$ kPa but also at higher pressure levels. This trend of behaviour of RREA is due seemingly to a higher co-ordination number (i.e., the number of inter-particle contact points per particle) associated with a high uniformity coefficient as well as an angular particle shape, resulting into a firm interlocking among stiff and strong core particles at the peak stress state. A firm interlocking among core particles may be hampered by the presence of thin mortar layers covering core particles, but it could be enhanced by high compaction and to some more extent by TC loading.

On the other hand, the E_{50} value of RREA increases with σ_c' at a much lower rate than q_{max}, reflecting the fact that particular effects of σ_c' on the $q - \varepsilon_v$ relation at small strains are not noticeable whereas the effects become evident only after the ε_v value exceeds a certain limit (see Fig. 7b). It may be seen that this limit strain becomes larger with an increase in σ_c'. This trend may be due likely to effects of the crushing of mortar layers covering core particles at inter-particle contact points during earlier stages of TC loading, resulting into insignificant effects of σ_c' on the E_{50} value. It is likely that, for the same reason, the behaviour of RREA becomes less dilative with an increase in σ_c' at a rate much higher than ordinary natural sand and gravel consisting of stiff and strong particles (Fig. 7b). On the other hand, it seems that the effects of the crushing of mortar layer become less significant as approaching the peak stress state, resulting into a significant increase in q_{max} with an increase in σ_c'.

3.3 Effects of compacted dry density and moulding water content

If the pre-peak stiffness and peak strength of crushed concrete aggregate are too sensitive to the water content during compaction, it is very difficult to use this material in actual construction projects, as a strict control of water content during compaction is often very difficult in the field. Moreover, it is necessary to know in advance the effects of compaction dry density, ρ_d, as well as those of compaction energy on the strength and deformation characteristics of a given backfill to determine the relevant compaction method in a given construction project. To evaluate the effects of these factors with crushed concrete aggregate, a set of CD TC tests ($\dot{\varepsilon}_v$=0.03%/min and $\sigma_c'=30$ kPa) were performed on REPA specimens compacted using three different energy levels ($E2'=5.06$ Nm/cm^3, $E1'=2.03$ Nm/cm^3 and $E0'=0.506$ Nm/cm^3) at different water contents, as shown in Fig. 9 (test series 3). Several other REPA specimens were also prepared by compacting to different dry densities at water contents, w, around w_{opt} (tests 04–09 & 13). To compare the sensitivity of the pre-peak stiffness and peak strength to ρ_d and w of crushed concrete aggregate with that of ordinary natural well-graded gravelly soils, model Chiba

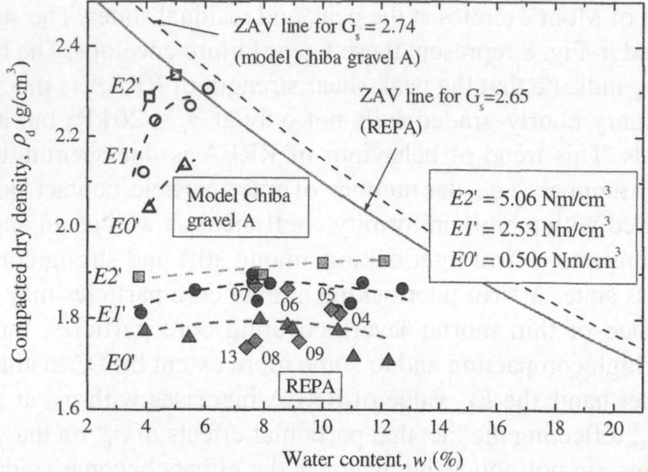

Figure 9. ρ_d–w relations of REPA and model Chiba gravel A compacted by three different energy levels and REPA compacted to different ρ_d values at w around w_{opt}, test series 3.

gravel A was prepared to have the same grading property as REPA (see Fig. 2 and Table 2). The $\rho_d - w$ relations of all the TC specimens used in test series 3 are presented in Fig. 9.

Figures 10a and b show the stress-strain relations of the specimens compacted at w around the respective w_{opt}. It may be seen that, for the same compaction energy and at the same σ_c', the pre-peak stiffness and compressive strength of REPA are generally higher than those of model Chiba gravel A. This result shows again that potential negative effects of having a thin mortar layer covering stiff and strong core particles can be largely alleviated by high-level compaction. On the other hand, with model Chiba gravel A, the initial stiffness does not increase noticeably with an increase in the compaction energy level from *E0'* to *E1'* while it decreases substantially with an increase from *E1'* to *E2'* (Fig. 10a). At small strains (Fig. 10b), the stiffness consistently decreases with an increase in the compaction energy level. Moreover, the volumetric strain characteristics become less dilative with an increase in the compaction energy level. These peculiar trends of behaviour are due likely to a more significant formation of thin horizontal cracks associated with vertical rebounding during compaction process of specimen in a compaction mould using larger compaction energy. However, these trends are not noticeable with REPA, if any.

Figures 11a and b show the relationships between q_{max} and the moulding water content, w, and between E_{50} and w. With REPA compacted at different water contents using the same energy level, there is a trend that the values of q_{max} and E_{50} become the respective maximum value when w is "w_{opt} for the respective compaction energy level". To examine whether this trend is due simply to the effects of changes in ρ_d by

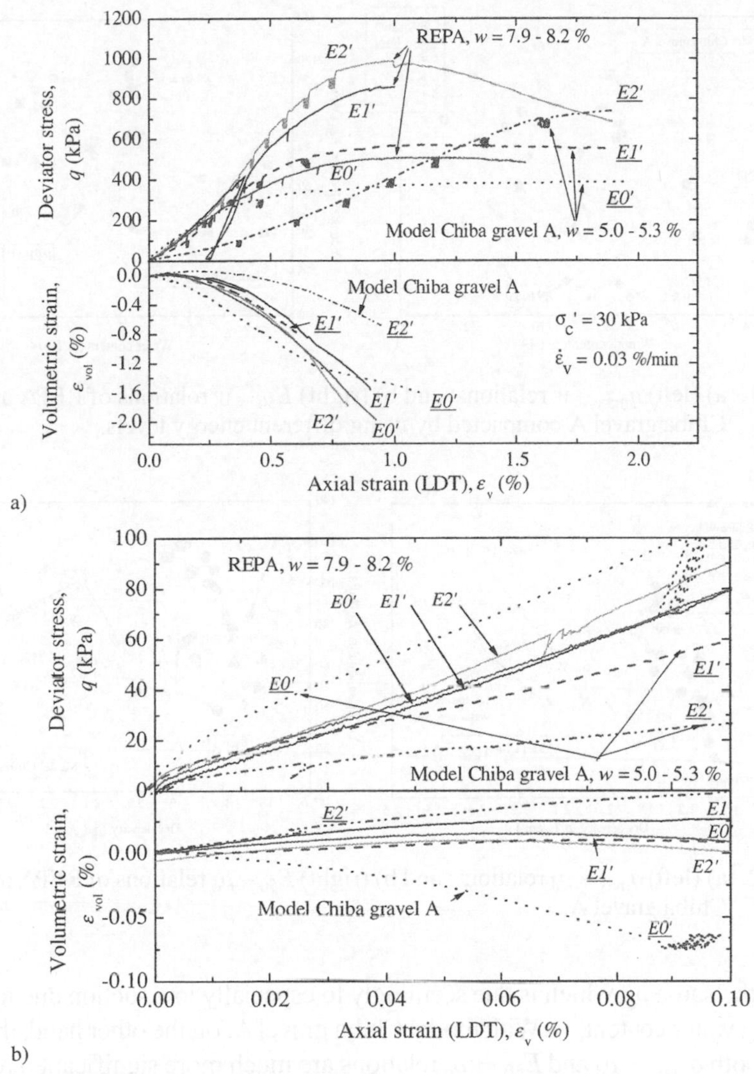

Figure 10. Stress-strain relations of REPA and model Chiba gravel A compacted by different energy levels at w around respective w_{opt}; a) large strain range; and b) small strain range, test series 3.

changes in w or whether additional effects of w by forming specific different microstructures at different water contents (e.g., Santucci de Magistris and Tatsuoka, 2003) are significant, the $q_{max} - \rho_d$ and $E_{50} - \rho_d$ relations for all the specimens were plotted (Figs.12a and b). It may be seen that the $q_{max} - \rho_d$ relation of REPA is rather unique, independent of w. It is also the case with the $E_{50} - \rho_d$ relation, although the relation is slightly affected by w. The q_{max} and E_{50} values of two specimens compacted at the driest state ($w = 3.8\%$) are noticeably lower than the respective average

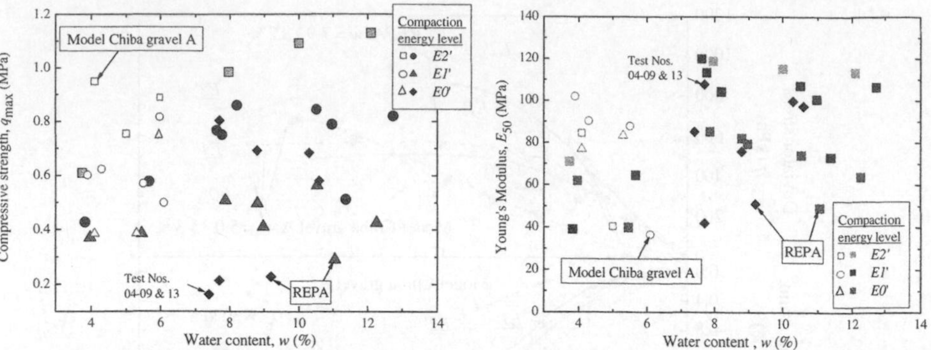

Figure 11. a) (left) $q_{max} - w$ relations; and b) (right) $E_{50} - w$ relations of REPA and model Chiba gravel A compacted by using different energy levels.

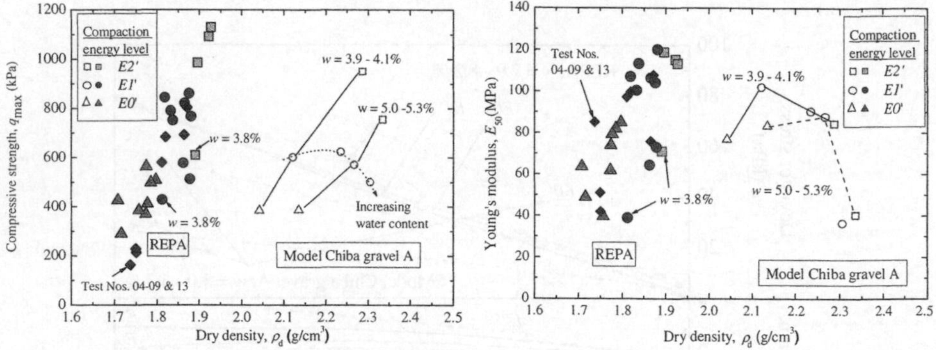

Figure 12. a) (left) $q_{max} - \rho_d$ relations; and b) (right) $E_{50} - \rho_d$ relations of REPA and model Chiba gravel A.

value at the same ρ_d, which is due seemingly to especially low suction due to too low moulding water content, w. With model Chiba gravel A, on the other hand, the effects of w on both $q_{max} - \rho_d$ and $E_{50} - \rho_d$ relations are much more significant than REPA. These results indicate that, as long as the compaction energy is high enough, the strict control of water content at compaction is less important with crushed concrete aggregate than with ordinary well-graded gravelly soil, which is another advantage of using crushed concrete aggregate as a backfill material.

It may also be seen from Fig. 12a that both q_{max} and E_{50} values of REPA are much more sensitive to ρ_d (i.e., larger slopes $dq_{max}/d\rho_d$ and $dE_{50}/d\rho_d$) than those of model Chiba gravel A. This result indicates that high compaction is inevitable to obtain sufficiently high pre-peak stiffness and peak strength particularly with crushed concrete aggregate.

The compaction of backfill is often controlled based on the degree of compaction, "$D_c = \rho_d$/" $\rho_{d.max}$ for a specified compaction energy level" or the compaction energy

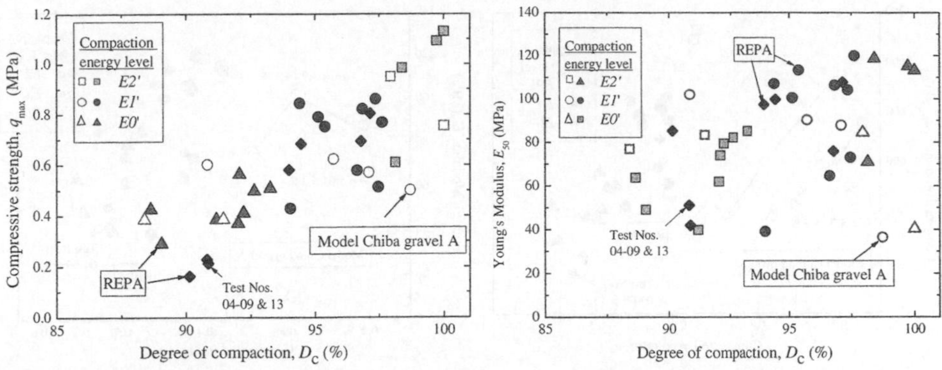

Figure 13. a) (left) $q_{max} - D_c$ relations; and b) (right) $E_{50} - D_c$ relations of REPA and model Chiba gravel A compacted by three different energy levels and REPA compacted to different ρ_d values at w around w_{opt}.

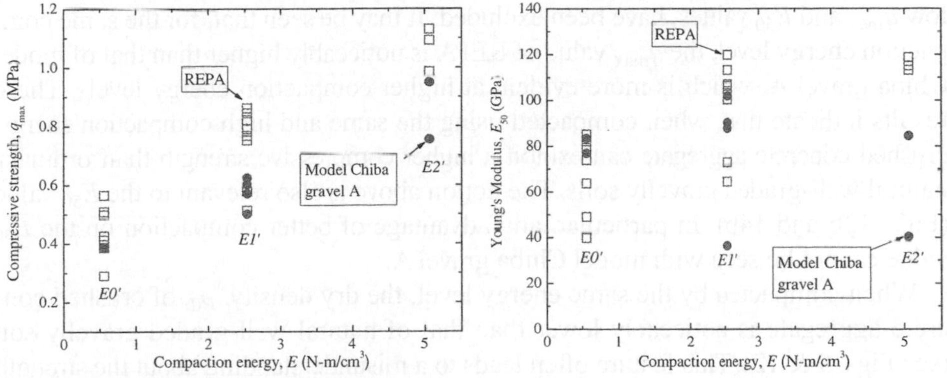

Figure 14. a) (left) q_{max} – compaction energy, E, relations; and b) (right) $E_{50} - E$ relations of REPA and model Chiba gravel A compacted at different water contents (two data of REPA compacted at $w = 3.8\%$ are excluded).

level or both. Figs. 13a and b compare the $q_{max} - D_c$ $(=\rho_d/\rho_{d.max})$ and $E_{50} - D_c$ relations of REPA and model Chiba gravel A, where $\rho_{d.max}$ is the respective maximum dry density for compaction energy level $E2'$. It may be seen that both q_{max} and E_{50} values of REPA are much more sensitive to the D_c value than those of model Chiba gravel A. In particular, the q_{max} value of REPA when the D_c value is 90%, which is often specified as the minimum required D_c value to be satisfied in a construction project, is only about 30% of the value when $D_c = 100\%$. This result indicates that the minimum requirement, $D_c = 90\%$, is too low with crushed concrete aggregate. When $D_c = 100\%$, the q_{max} value of REPA is noticeable higher than that of model Chiba gravel A (Fig. 13a). Fig. 14a compares the q_{max} values of REPA and model Chiba gravel A for three different compaction energy levels. In this plot, the data of the two REPA specimens compacted at $w = 3.8\%$, which exhibited exceptionally

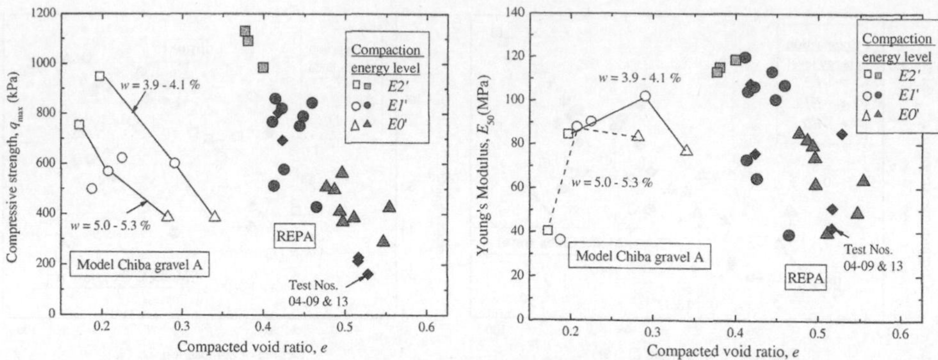

Figure 15. a) (left) $q_{max} - e$ relations; and b) (right) $E_{50} - e$ relations of REPA and model Chiba gravel A compacted by different energy levels.

low q_{max} and E_{50} values, have been excluded. It may be seen that, for the same compaction energy level, the q_{max} value of REPA is noticeably higher than that of model Chiba gravel A, which is more evident at higher compaction energy levels. These results indicate that, when compacted using the same and high compaction energy, crushed concrete aggregate can exhibit a higher compressive strength than ordinary natural well-graded gravelly soils. The notion above is also relevant to the E_{50} value (Figs. 13b and 14b). In particular, any advantage of better compaction on the E_{50} value cannot be seen with model Chiba gravel A.

When compacted by the same energy level, the dry density, ρ_d, of crushed concrete aggregate is noticeably lower than that of natural well-graded gravelly soil (see Figs. 9 & 12). This feature often leads to a misunderstanding about the strength and stiffness of crushed concrete aggregate. This relatively low ρ_d value of crushed concrete aggregate is due partly to a lower specific gravity, G_s, but mainly to higher compacted void ratios. It may be seen from Figs. 15a and b that REPA has generally much higher void ratios than a natural well-graded gravelly soil having the same grading characteristics and a similar grain shape. The void ratios of REPA (D_{max} = 19 mm) presented in Fig. 15 were obtained by using the G_s value of particles equal to and smaller than 10 mm. As the G_s value of crushed concrete aggregate generally decreases with a decrease in the particle size, these void ratios of REPA may be somehow under-estimated. It seems that this peculiar trend of behaviour (i.e., relatively high compressive strength despite relatively high void ratios) of crushed concrete aggregate is due to rather stable inter-particle contact of REPA resulting from its specific particle constitution (i.e., relatively soft and weak mortar layers covering stiff and strong core particles, Fig. 6), making the compaction more difficult while making the compressive strength higher.

It is to be noted that the effects of saturation under confined conditions on the strength and deformation characteristics of crushed concrete aggregate are negligible

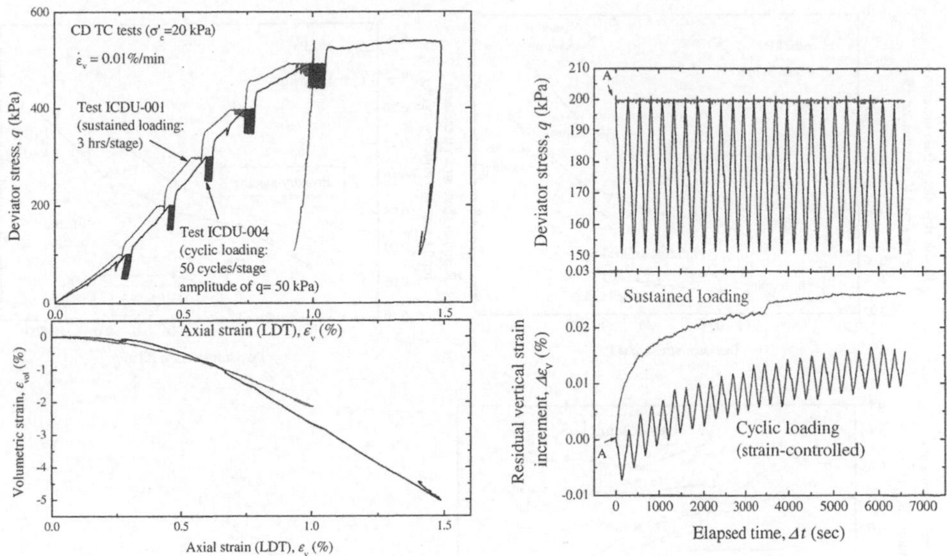

Figure 16. a) (left) Stress-strain relations from a pair of CD TC tests with sustained and cyclic loading stages; and b) (right) initial part of typical time histories of deviator stress and axial strain increment, RREA (test series 4, Aqil et al., 2005b).

(Aqil et al., 2005b), which is another important positive feature when used as the backfill material.

3.4 Residual strains by sustained and cyclic loading

The stiffness at small strains of crushed concrete aggregate is even smaller than that of poorly-graded granular materials (Fig. 5b) and does not increase noticeably with an increase in the confining pressure (Fig. 7b) and the compaction energy level (Fig. 10b). These defects indicate that crushed concrete aggregate may exhibit significant residual strains when subjected to long-term sustained and cyclic loading. To examine this feature, a series of drained TC tests were performed at $\sigma_c' = 20\,\text{kPa}$ on RREA compacted at w around w_{opt} using energy level *E1* (test series 4). Sustained loading for three hours or cyclic loading (50 cycles of deviator stress amplitude, $\Delta q = 50\,\text{kPa}$, at a constant strain rate) was applied during otherwise monotonic loading and unloading at an axial strain rate of $\pm 0.01\%$/min (test series 4). Figure 16 shows results from a typical pair of tests. Figure 17 shows the residual strain increments that took place for the first 110 minutes of sustained and cyclic loading, plotted against, respectively, the sustained deviator stress and the maximum or minimum deviator stress during cyclic loading applied during otherwise primary loading or global unloading. As seen from Figure 16, during primary loading, the sustained deviator stress and the maximum deviator stress during cyclic loading at the corresponding stages in the two

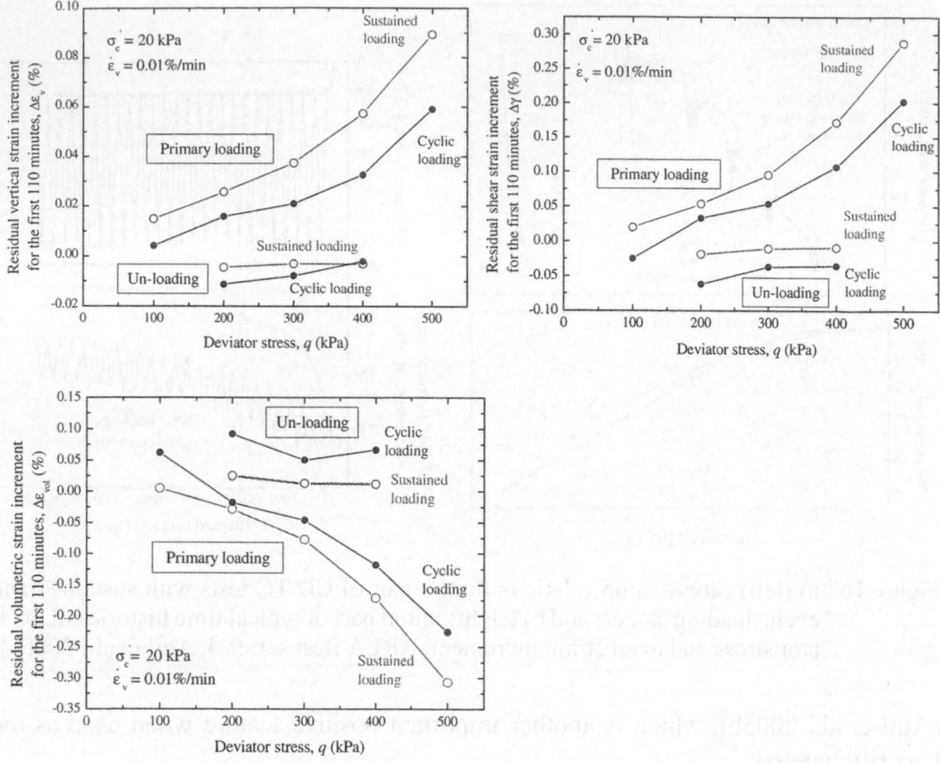

Figure 17. Comparison of residual vertical, shear and volumetric strain increments for the first 110 minutes of sustained and cyclic loading during otherwise primary loading and unloading, RREA.

tests were identical (Fig. 16). Figure 18 summarises the residual strain increments by full cyclic loading (50 cycles) at each stage. It may be seen from Figs. 17 and 18 that the residual strain increments of RREA by sustained and cyclic loading are not significant despite that it increases at a large rate with an increase in the shear stress level. It may also be seen from Fig. 17 that the residual axial and shear strain increments that develops by sustained loading is always larger than the value by cyclic loading under the loading conditions imposed in the present study.

The tangent stiffness during global unloading is substantially higher than the value during primary loading (Fig. 16a). Moreover, the residual vertical and shear strain increments caused by sustained and cyclic loading during otherwise global unloading are much smaller than those during the primary loading under otherwise the same conditions and the values are negative (Figs. 17 and 18; see also Fig. 19). These results indicate that relevant preloading, for example by compaction using a heavy machine, can decrease substantially the residual strain by sustained and cyclic

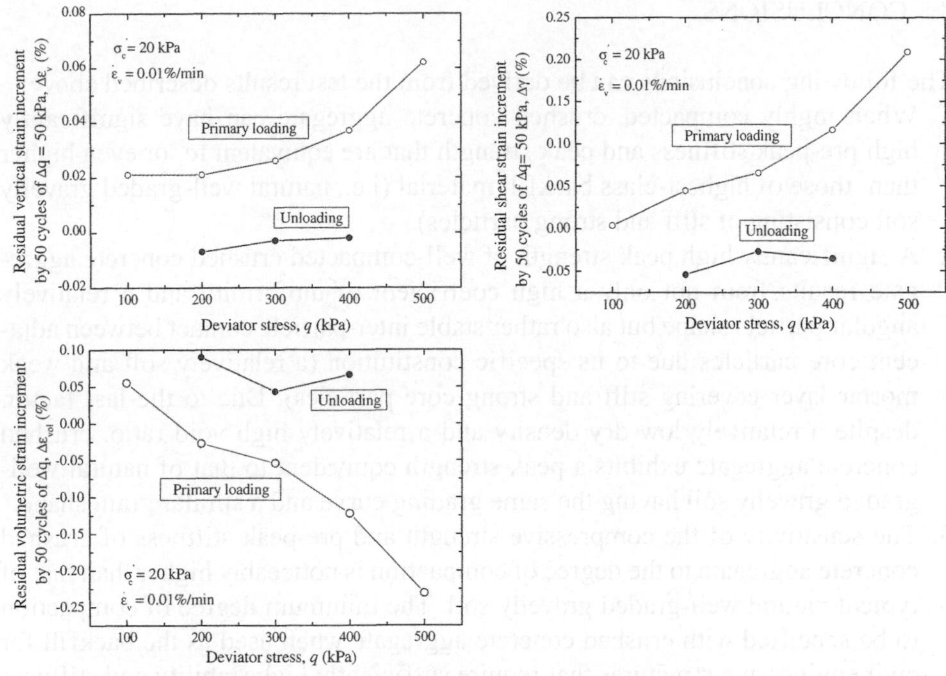

Figure 18. Comparison of residual vertical, shear and volumetric strain increments by 50 cycles of cyclic loading during otherwise primary loading and global unloading, RREA.

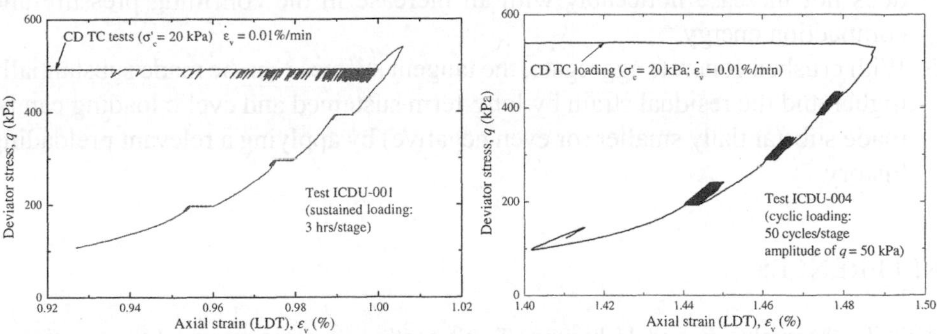

Figure 19. Deviator stress – axial strain relations during global unloading, a pair of CD TC tests described in Fig. 16a.

loading. The effects of preloading are generally higher with crushed concrete aggregate than with ordinary backfill (in particular poorly-graded ones). Tatsuoka et al. (1997) and Hasegawa and Shimakawa (2004) reported negligible long-term residual deformation of railway embankment constructed by highly compacting crushed concrete aggregate.

4 CONCLUSIONS

The following conclusions can be derived from the test results described above:

1. When highly compacted, crushed concrete aggregate can have significantly high pre-peak stiffness and peak strength that are equivalent to, or even higher than, those of highest-class backfill material (i.e., natural well-graded gravelly soil consisting of stiff and strong particles).
2. A significantly high peak strength of well-compacted crushed concrete aggregate results from not only a high coefficient of uniformity and a relatively angular particle shape but also rather stable inter-particle contact between adjacent core particles due to its specific constitution (a relatively soft and weak mortar layer covering stiff and strong core particles). Due to the last factor, despite a relatively low dry density and a relatively high void ratio, crushed concrete aggregate exhibits a peak strength equivalent to that of natural well-graded gravelly soil having the same grading curve and a similar grain shape.
3. The sensitivity of the compressive strength and pre-peak stiffness of crushed concrete aggregate to the degree of compaction is noticeably higher than that of typical natural well-graded gravelly soil. The minimum degree of compaction to be specified with crushed concrete aggregate when used as the backfill for civil engineering structures that require sufficiently high stability and stiffness should be much higher than 90%, say 95% or higher.
4. Due to the inclusion of mortar covering stiff and strong core particles, the initial stiffness at small strains of crushed concrete aggregate is relatively low and does not increase noticeably with an increase in the confining pressure and compaction energy.
5. With crushed concrete aggregate, the tangent stiffness can be made substantially higher and the residual strain by long-term sustained and cyclic loading can be made substantially smaller (or even negative) by applying a relevant preloading history.

REFERENCES

Aqil, U., Tatsuoka, F. and Uchimura, T., Strength and deformation characteristics of recycled concrete aggregate in triaxial compression, *Geo-Frontiers 2005 organized by Geosynthetic Materials Association, Geo-Institute of ASCE and Geosynthetic Research Institute at Austin, Texas, USA, January, Austin* (2005a).

Aqil, U., Tatsuoka, F., Uchimura, T., Lohani, T.N., Tomita, Y. and Matsushima, K., Strength and deformation characteristics of recycled concrete aggregate as a backfill material, *Soils and Foundations*, Vol. 45, No. 4, pp. 53–72 (2005b).

Goto, S., Tatsuoka, F., Shibuya, S., Kim, Y.-S. and Sato, T., A simple gauge for local strain measurements in the laboratory, *Soils and Foundations*, Vol. 31, No. 1, pp. 169–180 (1991).

Hasegawa, S. and Shimakawa, T., Use of recycled concrete aggregate to construct a subgrade for Tokaido Bullet Train Line for construction of a new Shinagawa Station, *the Foundation Engineering and Equipment (Kisoko)*, July, Vol. 32, No. 7, pp. 39–43 (2004) (in Japanese).

Mizukami, J., Kikuchi, Y. and Yoshino, H., Characteristics of concrete debris as rubble in marine construction, *Technical Note of the Port and Harbour Research Institute, Ministry of Transport, Japan*, No.906, June, p. 34 (1998) (in Japanese).

Santucci de Magistris, F. and Tatsuoka, F., Effects of molding water content on the stress-strain behavior of compacted silty sand, *Soils and Foundations*, Vol. 44, No. 2, pp. 85–102 (2003).

Sekine, E. and Ikeda, T., Geotechnical Characteristics of Crushed Concrete for Mechanical Stabilization, *Tsuchi-to-Kiso, Monthly Journal of Japanese Geotechnical Society*, May, Vol. 51, No. 5, pp. 31–33, (2003) (in Japanese).

Tatsuoka, F., Tateyama, M, Uchimura, T. and Koseki, J., Geosynthetic-Reinforced Soil Retaining Walls as Important Permanent Structures, 1996–1997 Mercer Lecture, *Geosynthetic International*, Vol. 4, No. 2, pp. 81–136 (1997).

11

Evaluation of Resilient Modulus and Environmental Suitability of Two Granular Recycled Materials

T.B. Edil
Department of Civil & Environmental Engineering, University of Wisconsin-Madison, U.S.A.

ABSTRACT: Resilient modulus of two recycled or processed materials (a foundry slag and a bottom ash from coal combustion) was determined in the laboratory on compacted specimens. The laboratory specimen moduli were then compared with those moduli back-calculated from large-scale laboratory proto-type pavement experiments and field falling weight deflectometer tests. The differences in stress and strain levels were taken into account and the necessary corrections were made to non-linear modulus to compare the moduli obtained by the different methods. Even when empirical corrections are applied, the elastic modulus obtained from a laboratory resilient modulus test tends to be lower than the operative elastic modulus obtained by back-calculation from prototype tests and field tests. For sand-like industrial byproducts, the field elastic moduli can be as much as four times higher than that measured in a laboratory resilient modulus test. The leachate quality of the two by-products shows that they discharge contaminants of concern at low concentrations immediately at the base of the by-product layer but the concentrations may exceed groundwater quality standards for some elements. However, this maximum concentration is expected to decrease with increasing depth to the groundwater table. The use of laboratory tests such as water leach test and column leach test may be insufficient to evaluate the impacts on groundwater from the use of industrial byproducts.

1 INTRODUCTION

The summary results of an investigation by the Pavement Geotechnics Group at The University of Wisconsin-Madison for development of recycled industrial by-products (referred also as processed materials) are presented. Resilient modulus is the basic material property required in mechanistic-empirical design of pavements. The test protocol for determining resilient modulus revolves around the cyclic triaxial testing of a soil specimen. Unlike many other mechanical properties of earthen materials, there is no direct way of verifying that the resilient modulus measured in the laboratory is representative of elastic moduli operative in the field under wheel loads. Therefore, there is uncertainty regarding how faithfully the resilient modulus measured in the laboratory corresponds to the operative elastic modulus in the field. Direct comparison of the resilient modulus with elastic moduli obtained from other methods is also difficult because of differences in stress and strain conditions, frequency of dynamic loading, and other factors. Additionally, the resilient behavior of industrial byproducts or other recycled materials introduced into pavement system is generally not well known as these materials are different than natural soils and also recently are being considered in large scale applications. For these reasons, the Pavement Geotechnics Group at The University of Wisconsin-Madison has been conducting a focused research on recycled materials in recent years. This paper summarizes certain aspects of this research; such as comparing the resilient modulus measured in the laboratory on test specimens of recycled or processed materials as well as a natural granular material with elastic moduli back calculated from a large-scale model experiment (LSME) and falling weight deflectometer (FWD) tests conducted in the field. The potential impact of such industrial by-products on groundwater was also evaluated using laboratory tests (e.g., water leach test and column tests) and in the field test sections using collection pan lysimeters. The observations relative to environmental suitability of these industrial by-products are also presented.

2 MATERIALS

A typical granular base course and two granular industrial by-products were used. Properties of the materials are summarized in Table 1. Grade 2 gravel is a natural material commonly used as base course in Wisconsin, U.S.A. Bottom ash (from a coal combustion power plant) and foundry slag (from a cupola furnace used for molten iron) are granular industrial by-products that are used as subbase over soft subgrades to enable construction. Both of the industrial byproducts are well-graded coarse-grained sand-like material.

Table 1. Properties of materials.

Material	Specific gravity	D_{10} (mm)	D_{60} (mm)	C_u	USCS symbol	Maximum dry unit weight (kN/m^3)		Optimum water content per D 698 (%)	CBR
						Compaction per ASTM D 698	Vibratory per ASTM D 4253		
Grade 2	2.65	0.09	6.0	66.7	GW	22.6	NM	8.2	NM
Bottom Ash	2.65	0.06	1.9	31.7	SW	15.1	13.7	–	21
Foundry Slag	2.29	0.13	2.0	15.4	SP	10.0	8.4	–	17

Note: NM = not measured.

3 MODULUS DETERMINATION METHODS

3.1 Laboratory resilient modulus test

Resilient modulus of the granular materials was evaluated using AASHTO T294-94 following the protocol for Type 1 materials (unbound granular base and subbase materials). The cell used for the resilient modulus test was identical to a triaxial cell used for shear strength testing, except air was used as the confining fluid instead of water. The load path consists of an axial, haversine load pulse 0.1 s in duration. This is followed by 0.9 s of material recovery for a total cycle time of 1 s. This cycle is intended to simulate the passing of one axle over a pavement followed by a period of rest before the next axle. It is repeated for the prescribed number of times in the standard while the applied load and deformation of the specimen are measured. This loading sequence is repeated 15 times at different values of confining pressure and deviator stress. The results are expressed in terms of resilient modulus (defined as the ratio of cyclic axial stress to elastic or recoverable component of axial strain) as a function of bulk stress (defined as the sum of principal stresses). There are several models proposed to represent the relationship of modulus to stress level. The following equation is widely used for granular materials:

$$M = k_1 \sigma_b^{k_2} \tag{1}$$

where k_1 and k_2 are empirical constants and σ_b is the bulk stress.

3.2 Large-scale model experiment (LSME)

The LSME is a method devised to model a pavement structure (or parts of it) at prototype scale in a manner that replicates field conditions as closely as practical (Tanyu et al. 2003). A schematic is shown in Fig. 1. A pavement profile is constructed

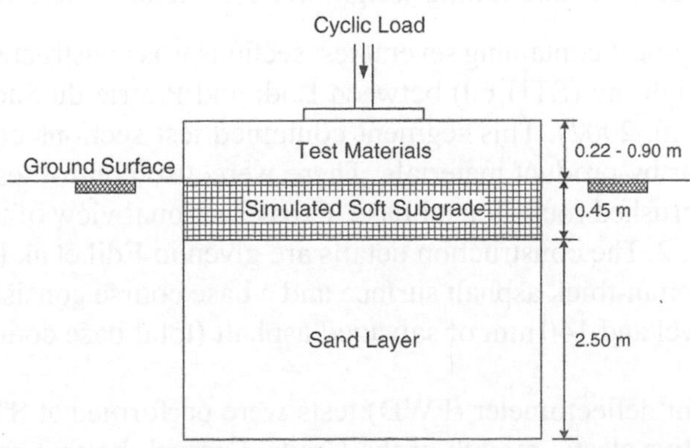

Figure 1. Schematic cross-section of large-scale model experiment (LSME).

in a 3 m × 3 m × 3 m test pit by placing three different layers of materials (from bottom to top): (i) a 2.5-m thick layer of dense uniform sand, (ii) a 0.45-m-thick simulated soft subgrade (expanded polystyrene foam), and (iii) a layer of coarse granular test material (0.22 to 0.90-m thick) simulating a working platform placed to enable construction on a soft grade, which acts subsequently similar to a subbase in the finished pavement system. A riding surface and a base course layer are not incorporated in the LSME, but their effect in transmitting wheel loads is considered in the loads applied. Repetitive loads are applied to the surface of the profile with a steel plate using a 90-kN hydraulic actuator. A multi-layer elastic analysis indicated that stress applied to the subbase layer is typically about 140 kPa, or approximately 20% of that applied to the surface of the pavement. This stress was simulated by applying a force of 7 kN to the loading plate (with a diameter of 250 mm).

A multilayer elastic analysis program, KENLAYER (Huang 1993), was used to invert the elastic modulus of the subbase layer in the LSME from the measured loads and defections (Tanyu et al. 2003). The simulated soft subgrade was assumed to be linearly elastic, whereas the elastic modulus of the subbase was assumed to follow the elastic power function in Eq. 1. The subbase was divided into sublayers 50-mm thick in the analysis.

The parameter k_2 varies in a narrow range for a wide variety of granular materials. Thus, k_2 was fixed using the value obtained from the resilient modulus test conducted per T294. The parameter k_1, which varies over a broad range, was adjusted until the measured and predicted elastic deflections matched. The k_1 that provided matching deflections was assumed to be the operative k_1 of the subbase layer. The elastic deflections used as input to KENLAYER were derived from the total deflections (elastic and plastic) measured in the LSME by subtracting the accumulated plastic (non-recoverable) deflections from the total deflections.

3.3 Field test sections and falling weight deflectometer tests (FWD)

A 654-m test segment containing several test sections was constructed during reconstruction of a highway (STH 60) between Lodi and Prairie du Sac in Wisconsin, U.S.A. (Edil et al. 2002). This segment contained test sections constructed with the two granular by-product materials. There were also control test sections constructed using crushed rocks for subbase. Cross-sectional view of the test sections is shown in Fig. 2. The construction details are given in Edil et al. (2002). All sections had a 125-mm-thick asphalt surface and a base course consisting of 115 mm of Grade 2 gravel and 140 mm of salvaged asphalt (total base course thickness = 255 mm).

Falling weight deflectometer (FWD) tests were performed at STH 60 to determine the operative elastic moduli of the Grade 2 gravel, bottom ash, and foundry

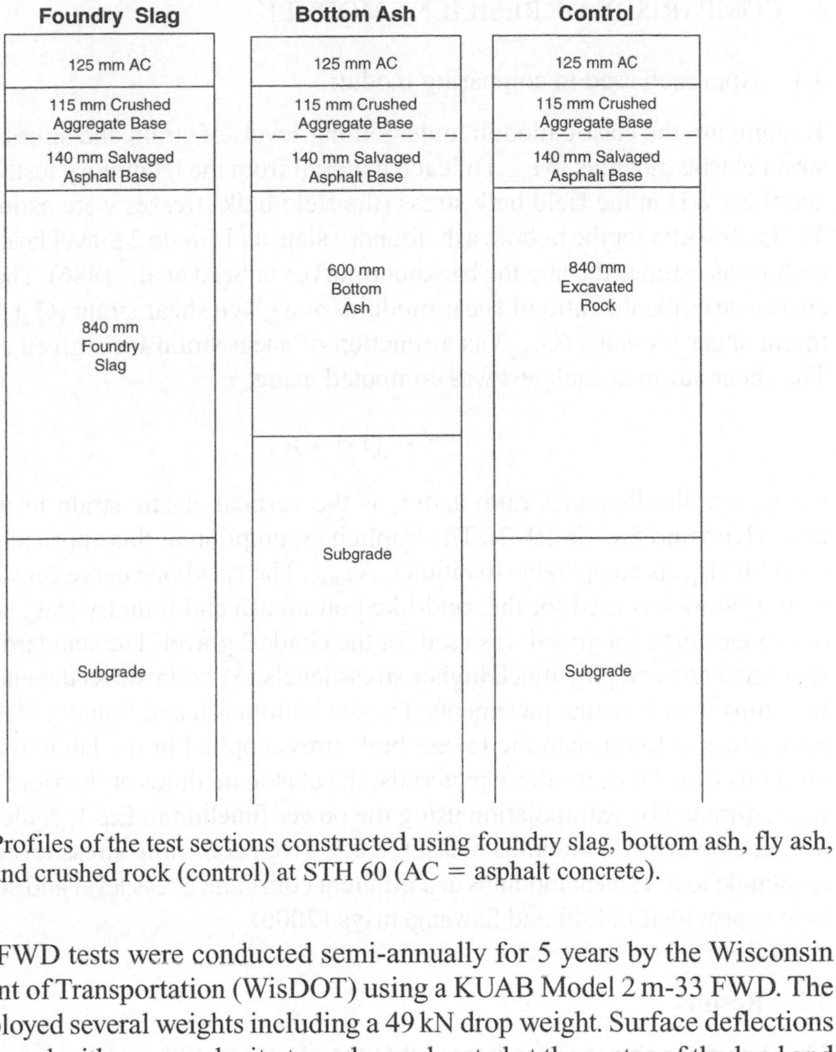

Figure 2. Profiles of the test sections constructed using foundry slag, bottom ash, fly ash, and crushed rock (control) at STH 60 (AC = asphalt concrete).

slag. The FWD tests were conducted semi-annually for 5 years by the Wisconsin Department of Transportation (WisDOT) using a KUAB Model 2 m-33 FWD. The FWD employed several weights including a 49 kN drop weight. Surface deflections were measured with seven velocity transducers located at the center of the load and at 6 different distances from the center of the load.

Elastic moduli of the subbase and base layers were computed using the loads and deflections measured with the FWD. The layered elastic analysis program MODULUS (Texas Transportation Institute 1991) was used for the analysis because it provides routines for back-calculating elastic moduli from FWD data. Each pavement layer was assumed to be homogeneous, isotropic, and linearly elastic and to extend infinitely in the horizontal direction. The bottom layer is also assumed to extend downward infinitely. Elastic moduli assigned to each layer are adjusted iteratively until the measured and predicted deflections match within an accepted tolerance (Uzan et al. 1988).

4 COMPARISON OF RESILIENT MODULI

4.1 Approach used in comparing moduli

To compare the elastic moduli under similar levels of strain and stress, the small-strain elastic modulus (M_{max}) of each material from the laboratory test, the LSME, and the FWD at the field bulk stress (the field bulk stresses were estimated to be 31, 35, 104 kPa for the bottom ash, foundry slag, and Grade 2 gravel layers, respectively) was estimated using the backbone curves in Seed et al. (1986). The backbone curves describe the ratio of shear modulus at a given shear strain (G_γ) to the maximum shear modulus (G_{max}) as a function of shear strain for a given stress level. The shear strain in each test was computed using:

$$\gamma = (1 + \upsilon)\varepsilon_v \tag{2}$$

where υ is the Poisson's ratio and ε_v is the vertical elastic strain in the subbase layer (Kim and Stokoe 1992). The implicit assumption in this approach is that the ratio M/M_{max} is comparable to ratio $G_{\gamma'}/G_{max}$. The backbone curve for sand in Seed et al. (1986) was used for the sand-like bottom ash and foundry slag, whereas the backbone curve for gravel was used for the Grade 2 gravel. The standard laboratory specimen tests employ much higher stress levels to obtain stress dependency relationships than in actual pavements. For the bottom ash and foundry slag, the field bulk stress is lower than the lowest bulk stress applied in the laboratory resilient modulus test. Thus, for these materials, the elastic modulus at the field bulk stress was estimated by extrapolation using the power function in Eq. 1. A clear description of converting a modulus measured at a given confining stress level and strain amplitude to a resilient modulus at a different confining stress level and strain amplitude is provided in Edil and Sawangsuriya (2006).

4.2 Results

The data from this study indicated that the elastic modulus of granular materials, including granular industrial by-products, used in pavement systems depends on the state of stress and the strain amplitude. Comparison of elastic moduli measured in a conventional laboratory test to those operative at the prototype and field scales requires that state of stress and strain amplitude be comparable. Differences in strain amplitude can be dealt with by applying an empirical correction using backbone curves for granular materials. The effects of stress are handled by limiting comparisons only to elastic moduli measured at a comparable state of stress. However, the laboratory resilient modulus test does not always provide elastic moduli for the range of bulk stresses encountered in the field. In such cases, an estimate of the elastic modulus at the field state of stress can be obtained by extrapolation using a power function.

Table 2. Low-strain modulus obtained at field bulk stress.

Method	Elastic modulus M (MPa)	Shear strain γ (%)	Shear modulus ratio M/M_{max}	Maximum modulus M_{max} (MPa)
Bottom Ash, Bulk Stress = 31 kPa				
FWD	108	0.038	0.48	225
LSME (0.46 m)	28	0.175	0.18	156
LSME (0.92 m)	72	0.045	0.46	157
Lab Test	32	0.018	0.60	53
Foundry Slag, Bulk Stress = 35 kPa				
FWD	119	0.029	0.57	209
LSME (0.46 m)	20	0.352	0.15	133
LSME (0.92 m)	29	0.173	0.23	130
Lab Test	30	0.029	0.57	53
Grade 2 Gravel, Bulk Stress = 104 kPa				
FWD	143	0.049	0.30	477
LSME (0.23 m)	28	1.001	0.07	400
LSME (0.46 m)	68	0.075	0.23	296
Lab Test	157	0.018	0.43	365

Even when corrections are applied, the small-strain elastic modulus (M_{max}) obtained from a laboratory resilient modulus test tends to be lower than the operative elastic modulus obtained by back-calculation from prototype tests (e.g., the LSME) and field tests using the FWD (Table 2). For conventional gravels used as subbase materials, the difference between the elastic moduli measured in the laboratory and field is small. However, for sand-like industrial byproducts, the field elastic moduli can be as much as four times higher than that measured in a laboratory resilient modulus test. Tanyu et al. (2006) presented design charts that show the structural number or the roadbed modulus as a function of type and thickness of the by-product layer.

5 FIELD LYSIMETERS

Pan lysimeters (3.50 m × 4.75 m in area) were installed beneath each test section constructed with industrial by-products as well as the control section constructed of crushed rock in STH 60 field test segment (Edil et al. 2002). The lysimeters were installed to determine the amount of liquid passing through the pavement structure at the base of the by-products layer and to determine the concentration of select contaminants (cadmium, chromium, selenium, and silver) in the leachate. A typical layout and construction of the lysimeters is shown in Fig. 3.

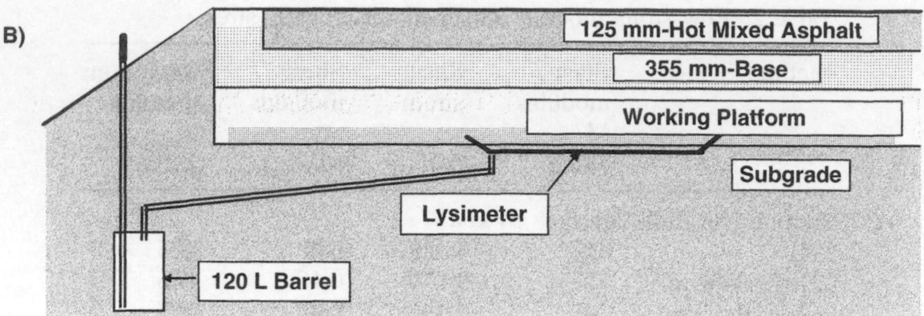

Figure 3. Cross-section of lysimeters located at STH 60 (working platform refers to by-product layer)

The pan lysimeters were constructed with 1.5-mm thick textured linear low density polyethylene geomembrane overlain by a geocomposite drainage layer comprised of a geonet with nonwoven needle-punched geotextiles heat bonded to either side. Each lysimeter drains to a 120-L polyethylene collection tank buried in the shoulder. The collection tanks are insulated with extruded polystyrene to prevent freezing. Leachate that accumulates in the collection tanks is removed on a regular basis using a pump for chemical analysis.

6 ENVIRONMENTAL SUITABILITY

Once the highway was paved during the first week of October 2000, the fluxes decreased dramatically. The flux during the winter was very small because of frozen conditions. In spring, the leachate fluxes increased to values comparable to those measured in October 2000 before frost penetrated the pavement. The average fluxes from the foundry slag (0.22 mm/d) and control (0.20 mm/d) sections were not as high as the average flux from the bottom ash section (0.26 mm/d).

Groundwater quality standards applicable to the field site are defined in Section NR 140 of Wisconsin Administrative Code, *Groundwater Quality NR 140*. The standards in NR 140 are similar to, or lower than USEPA maximum contaminant levels (MCLs). A comparison of the NR 140 standards for Cd, Cr, Se, and Ag and peak concentrations from the test sections is given in Table 3.

Cd concentrations in the leachate from the foundry slag, bottom ash, and control sections exceeded the NR 140 standard (5 μg/L) by a factor of 4–6 in the field test. However, in all cases, the Cd concentrations were below the NR 140 standard after 16 mos and 0.6 PVF. Se concentrations exceeded the NR 140 standard (50 μg/L) for the by-product test sections by a factor of about 1.5. Moreover, in all cases, the Se concentration increased and then leveled off at a concentration exceeding the NR 140 standard over the last 24 months of monitoring. In contrast to Cd

Table 3. Peak lysimeter concentrations observed in field tests.

Material	Peak lysimeter conc. ($\mu g/L$)			
	Cd	Cr	Se	Ag
Foundry Slag	32.1	49.6	151	8.2
Bottom Ash	21.2	32.1	141	15.2
Control	6.2	3.3	100	3.9
NR 140 Requirements	5	100	50	50

and Se, none of the test sections had Cr or Ag concentrations exceeding the NR 140 standard (100 μg/L for Cr and 50 μg/L for Ag). Leachate collected in the lysimeters is representative of pore fluid at the bottom of the pavement profile and represents water reaching groundwater only if the groundwater table is at the base of the pavement profile. In many roadways, the water table will be deeper. Processes such as sorption, diffusion, dispersion, and dilution occurring in soils between the base of the pavement and the groundwater table will likely to result in lower concentrations by the time the groundwater table is reached. Bin Shafique et al. (2002) conducted a modeling study to simulate the transport of contaminants from working platforms constructed with by-products to the groundwater table using a validated model of flow and transport. Their findings indicate that the maximum relative concentration decreases with increasing depth to the groundwater table. In particular, the maximum concentration 1 m below the pavement layer typically was 20% of the peak concentration at base of the byproduct layer and 10% of the peak concentration 5 m below the byproduct layer.

The re-use of industrial byproducts in Wisconsin is regulated based on water leach test (WLT) concentrations. WLTs were conducted by Sauer et al. (2005) on foundry slag and bottom ash using the methods in ASTM D 3987. The bottom ash and foundry slag used in the study essentially met the Wisconsin requirements for use in confined geotechnical fills. The WLTs, however, did not produce leachate with Cd, Cr, Se, and Ag concentrations that were consistent with the leachate concentrations observed in the field. WLTs may be an insufficient index test to characterize the potential for leaching of metals from industrial byproducts in the field.

Similarly, Sauer et al. (2005) conducted column leach tests (CLTs) as an alternative to WLTs, with the thought that the flow-through conditions of the CLTs more closely represent the conditions in the field. CLTs on the granular materials (foundry slag and bottom ash) were conducted using rigid-wall permeameters. Specimens were compacted directly into a PVC column having the same size as a standard Proctor mold (101.6 mm in diameter and 114.3 mm tall). Upward flow was imposed using a peristaltic pump set at 7 mL/hr (2 mm/d) to 30 mL/hr (9 mm/d). A 0. 1 M LiBr solution was used as the influent in the column tests simulating the ionic strength of the pore water in pavement layers. Volume and pH of the effluent were recorded each time an effluent sample was collected for chemical analysis. The concentrations in the effluent from these CLTs did not closely simulate the concentrations found in the leachate from the field tests. The use of laboratory tests such as WLTs and CLTs may be insufficient to evaluate the impacts on groundwater from the use of industrial byproducts. A systematic approach incorporating the physical and chemical properties of the material and the hydrogeological features of the area should be implemented before a byproduct is used.

7 SUMMARY AND CONCLUSIONS

The elastic modulus of granular materials, including granular industrial by-products, used in pavement systems depends on the state of stress and the strain amplitude. Comparison of elastic moduli measured in a conventional laboratory test to those operative at the prototype and field scales requires that state of stress and strain amplitude be comparable. Even when empirical corrections are applied, the elastic modulus obtained from a laboratory resilient modulus test tends to be lower than the operative elastic modulus obtained by back-calculation from prototype tests (e.g., the LSME) and field tests using the FWD. For conventional gravels used as subbase materials, the difference between the elastic moduli measured in the laboratory and field is small. However, for sand-like industrial byproducts, the field elastic moduli can be as much as four times higher than that measured in a laboratory resilient modulus test.

The leachate quality observations of the by-products in the field shows that they discharge contaminants of concern at low concentrations immediately at the base of the by-product layer but the concentrations may exceed groundwater quality standards for some elements. However, this maximum concentration is expected to decrease with increasing depth to the groundwater table. The use of laboratory tests such as WLTs and CLTs may be insufficient to evaluate the impacts on ground water from the use of industrial byproducts.

ACKNOWLEDGEMENT

The research summarized here was conducted by the author and his colleague Dr. C.H. Benson and former graduate students, Drs. B.F. Tanyu and W.H. Kim, and Mr. J.J. Sauer over the last 5 years. The material for this paper was gathered largely for a presentation in TC-3 Geotechnics of Pavements Workshop in Prague. Financial support for the research was provided by the Wisconsin Highway Research Program for the mechanical segment of the study and the field tests. The Recycled Materials

Research Center through the Wisconsin Department of Transportation, the Wisconsin Department of Natural Resources Waste Reduction, and the Recycling Demonstration Grant Program through Alliant Energy provided support for the environmental impact studies.

REFERENCES

Bin Shafique, S., Benson, C.H., & Edil, T.B. 2002. *Leaching of heavy metals from fly ash stabilized soils used in highway pavements*. Geo Engineering Report No. 02-14, Department of Civil and Environmental Engineering, University of Wisconsin-Madison.

Edil, T.B., Benson, C., Bin-Shafique, M., Tanyu, B., Kim, W. & Senol, A. 2002. Field evaluation of construction alternatives for roadway over soft subgrade. Transportation Research Record, 1786, Transportation Research Board, National Research Council, Washington, DC: 36–48.

Edil, T.B. & Sawangsuriya, A. 2006. Use of stiffness and strength for earthwork quality evaluation. *GeoShanghai International Conference*. Shanghai, China (in press).

Huang, Y.H. 1993. *Pavement Analysis and Design*, Prentice Hall, Inc., Englewood Cliffs, New Jersey.

Kim, D.S. & Stokoe II, K.H. 1992. Characterization of resilient modulus of compacted subgrade soils using resonant column and torsional shear tests. *Transportation Research Record*, 1369, Transportation Research Board, National Research Council, Washington, DC: 83–91.

Sauer, J.J., Benson, C.H. & Edil, T.B. 2005. *Metals leaching from highway test sections constructed with industrial by-products*. Geo Engineering Report No. 05-21, Department of Civil & Environmental Engineering, University of Wisconsin-Madison, Wisconsin.

Seed, H.B., Wong, R.T., Idriss, I.M. & Tokimatsu, K. 1986. Moduli and damping factors for dynamic analyses of cohesive soils, *Journal of Geotechnical Engineering*, ASCE, 112 (11): 1016–1032.

Tanyu, B.F., Kim, W.H., Edil, T.B. & Benson, C.H. 2003. Comparison of laboratory resilient modulus with back-calculated elastic moduli from large-scale model experiments and FWD tests on granular materials. *Resilient Modulus Testing for Pavement Components*, ASTM STP 1437, G.N. Durham, A.W. Marr, and W.L. DeGroff, Eds., ASTM International, West Conshohocken, PA.: 191–208.

Tanyu, B.F., Kim, W.H., Edil, T.B. & Benson, C.H. 2006. Development of methodology to include structural contribution of alternative working platforms in a pavement structure. *Transportation Research Record*, Paper No. 05-1395, Transportation Research Board, Washington, DC, (in press).

Texas Transportation Institute 1991. *MODULUS 4.0, User's Manual*, Texas Transportation Institute, Texas A&M University, College Station, TX, USA.

Uzan, J., Scullion, T., Michalak, C.H., Paredes, M. & Lytton, R.L. 1988. *A microcomputer based procedure for backcalculating layer moduli from FWD data*. Research Report No. 1123-1, Texas Transportation Institution, College Station, Texas, USA.

12

Current State of the Use of Recycled Materials in Geotechnical Works in Portugal

E. Fortunato & F. Pardo de Santayana
Laboratório Nacional de Engenharia Civil, Lisbon, Portugal

ABSTRACT: The reuse of recycled materials in geotechnical works has awakened a growing interest in the last few years in Portugal. A committee (GT-VROG) within the Portuguese Geotechnical Society (SPG) was created recently for analysing the current state of this kind of use and for promoting it nationwide. As part of the activities of the group, a national survey within producers of waste materials and potential users in geotechnical works was carried out, and a national seminar was organised within the framework of the activities of the TC-3 of the ISSMGE, in order to discuss the present situation in the country and the best ways to develop the reuse of this kind of materials in geotechnical applications. The seminar had the participation of representatives of the different entities interested on the subject (producers, users, administration and research centres), as well as of invited foreign specialists from other European countries. Taking advantage of the results of these activities, the current state of the use of recycled materials in geotechnical works in Portugal is briefly presented and analysed next.

1 INTRODUCTION

Currently, the reuse of waste materials in civil engineering works rather than an option should be an imperative aspect. This field of activity should be adapted to the principles of sustainable development, by reducing the exploitation of natural non-renewable materials, and by making available the space that would be occupied by recyclable waste in landfills. In fact, many millions of cubic metres of natural aggregates and soils are used every year in pavements and transportation earthworks, for instance, amounts that could be significantly reduced by using non-traditional recycled materials. Nevertheless, under no circumstance the quality and the economic rationality of civil engineering projects should be affected by the use of materials and techniques different from those that are established.

In Portugal, there has been some effort in studying the possibility of using recycled materials in civil engineering works, namely in geotechnical works. The present work refers to the difficulties usually encountered in implementing a policy of promoting the use of recycled materials. Mention is made to the main requirements of any programme for developing the use of recycled materials in geotechnical applications, and to the ways to overcome the difficulties in developing their use, as well as to the main agents that must be involved in implementing that policy. In addition, and as a result of the activities of the committee GT-VROG, of the Portuguese Geotechnical Society (SPG), devoted to the reuse of waste in geotechnical works, a few studies and recent applications developed in Portugal in the field of recycled materials in pavements and transportation earthworks are presented, namely relatively to: end-of-life tyres; municipal solid waste incinerator bottom ash; water treatment plant sludge; slag and ashes from thermal power plants; quarry wastes (gravel and crushed rocks; quarry mud); steel slag; and construction and demolition waste.

2 SOME CONSIDERATIONS ABOUT THE USE OF RECYCLED MATERIALS

Recycling waste materials in civil engineering works contributes to adapt construction activities to the principles of sustainable development, therefore reducing the exploitation of natural, non-renewable materials, and making available the space that would be occupied by landfills. Several million cubic metres of natural aggregates and soils are annually used in pavement and transportation earthworks, for instance. These quantities could be significantly reduced by resorting to the use of non-traditional materials.

The essential factors to be taken into account for implementing a waste reuse policy should be as follows: civil engineering issues, namely as regards characterisation of materials; geo-environmental factors; and economic factors.

The main requirements of a waste reuse programme in geotechnical applications could be summarized as follows: identifying the application; defining the main requirements of the application; verifying the environmental appropriateness; developing laboratory characterisation tests; modelling the geotechnical performance; preparing construction specifications; observing workability and "in situ" performance; assessing the long term performance; and disseminating the technical and scientific knowledge.

The main agents that are to be involved in a waste reuse policy are the following: producers; potential users; designers and contractors; transport institutions; state bodies/ministries; research/advisory centres; and associations for the environment.

The main difficulties that are normally found in implementing a waste reuse policy are as follows: lack of appropriate information on the short and long-term performance of materials; lack of standardised test methods; lack of information on the most appropriate construction methods; insufficiency of available materials; problems related with the homogeneity of the material; high transportation costs comparatively with traditional materials; lack of teams, time and financial means to develop legislation, to conduct research works and to give appropriate training; disconnected policies at the different decision levels.

The solutions to overcome waste reuse difficulties should be based on: increasing research and demonstration projects; increasing transfer of knowledge among bodies and among countries; increasing the use of financial supports or incentives; producing effective legislation; convincing sceptics; and defining and promoting the use of waste in contracts.

Taking into account the characteristics of pavement and transportation earthworks, as well as the current concerns in terms of economy, environment and performance, it is necessary to become familiar with the details of all aspects that contribute to the high quality of projects. With regard to the characterisation of recycled materials, this usually begins by the determination of their chemical and mineralogical properties, taking into account the fact that the materials will be used to build different pavement layers and will be subject to rainwater infiltration and seepage. For this reason, it is necessary to determine if the materials include potentially harmful soluble elements, which will contribute to contaminate the ground water and soils.

Knowing the physical characteristics of materials is also necessary in order to estimate the parameters to be used for designing the structures. Comparison of these parameters with those presented by traditional materials is always a useful means to evaluate the suitability of a particular recycled material. However, construction specifications are usually based on the performance of traditional materials. In the absence of specific requirements for the recycled materials, tests are usually performed in a similar way as for the traditional materials. For this reason, the results of these tests have to be interpreted with caution. Actually, the suitability of the materials for geotechnical applications has to be evaluated by means of appropriate tests, which enable to know both the short and the long-term behaviour of these materials when subject to conditions representative of those imposed on during construction and operation.

Aspects related with the application of materials on the site, namely transportation, and with the workability of materials and construction methods, should be the object of studies conducted under conditions similar to the real ones. Very often, these are the aspects that condition the application. Even though the user's guides contain very useful recommendations, frequently they are too focused on local materials, being necessary to take into account the differences in the characteristics of the same type of wastes but coming from different countries or regions.

It is felt that, besides an adequate laboratory characterisation of the recycled materials to be used, it is always necessary to promote field trials, with characteristics as similar as possible to those of the works to be carried out, and to control these trial sections by means of performance based specifications. In addition to this, it is always necessary to specify an observation plan to enable the monitoring of the behaviour of the work during the different stages. Only by adequately monitoring the experiments performed and by disseminating the results obtained, it is possible to mobilise the technical and scientific communities and to convey confidence to the authorities, so that they may promote the application of recycled materials.

3 SOME ASPECTS OF THE ACTIVITY RECENTLY DEVELOPED IN PORTUGAL

A Portuguese committee was created by the Geotechnical Portuguese Society for promoting the use of recycled materials in geotechnical applications, involving researchers of different institutions and representatives of public and private entities related to the production or use of recycled materials in geotechnical applications (roads and railways authorities, governmental agencies and industry agencies). The main objectives, which led to the creation of this group, were the following: to gather information and to analyse the current state as regards promoting the use of recycled materials in geotechnical works in Portugal; and to disseminate and promote the use of these materials. Two main actions have been undertaken by this committee:

i) a questionnaire was elaborated and circulated by the technical and scientific communities at national level; the main objective of this questionnaire was to compile available information about: main characteristics of wastes materials produced in Portugal; type of applications that have been made of these recycled materials; informative knowledge of producers and users; difficulties experienced by these entities to use waste materials; and developments expected for the use of waste materials;
ii) a seminar was organised in order to discuss the most relevant aspects within the scope of the use of recycled materials in geotechnical works in Portugal; the objectives of the seminar were to disseminate the state-of-the-art of the application of these materials in various countries, to identify the different Portuguese experiences, and to provide and encourage, in the near future, the use of recycled materials in the country.

4 GT-VROG QUESTIONNAIRE

Within the framework of the actions developed by the GT-VROG, a nation-wide questionnaire was carried out about the current state of the use of waste in geotechnical works in Portugal. This questionnaire was intended to provide some knowledge about the main characteristics of the wastes produced in Portugal, the use that has been done of those wastes, the level of knowledge of producers and of potential users, the difficulties experienced by those agents as regards this issue, and their expectations concerning the evolution of the present situation. The dissemination of the questionnaire began in 2003. To date, some difficulties have been encountered in obtaining responses, which may indicate a certain lack of information on the part of those to whom the questionnaires were sent, as regards the issues addressed. The initial conclusions, which could be drawn from the response to the questionnaire, are as follows:

i) there is some waste that is usually used by being incorporated into cement, such as for example, sludge resulting from water treatment;
ii) some producers have already promoted the chemical, physical and mechanical characterisation of the recycled materials produced, and have made available the values obtained;
iii) those producers have little knowledge about national and international legislation related to the use of recycled materials in geotechnical works;
iv) various producers of waste have an institutional environmental management system in place, certified by the appropriate authorities;
v) the main obstacles to the use of the recycled materials are related to: the lack of guidelines regarding the reuse of recycled materials; technical and economic difficulties as to the treatment of waste prior to being used and to its transport; lack of knowledge regarding the behaviour of recycled materials when incorporated into structures; inadequate legislation; lack of interest by government authorities and public companies; difficulties in sharing responsibilities; lack of a map of recycled materials;
vi) generally speaking, the entities that responded to the questionnaire showed an interest and availability in seeking solutions for the final destination of waste, albeit without specifying the type of participation they would be willing to take. Potential users stressed the need to carry out field studies and field trials to enable the validation of options for the use of recycled materials;
vii) although, in general, recycled materials are not systematically used in geotechnical works, there are some specific cases of studies for their application in pavement and transportation earthworks, for example: a) end-of-life tyres; b) municipal solid waste incinerator slags; c) water treatment plant sludges; d) slags and ashes from thermal power plants; e) waste gravel and crushed rocks resulting from quarrying.

5 A FEW ASPECTS APPROACHED IN THE SEMINAR

General approaches to the integrated waste management policy and to the legal aspects related with the reuse of waste in Portugal were presented in articles devoted to this issue (Lobo and Santiago, 2004; Oliveira, 2004). It has been stressed that the hierarchy of actions in waste management has been inverted. In fact, up to now, the main practice has been the deposition in waste landfills. Such practice is part of the Final Elimination, in detriment to the activities of prevention, which are intended to reduce the amount of waste, as well as the cost associated with treatment and the corresponding environmental impact, and, also, in detriment to the activities intended to take advantage of the materials (reuse, recycling or energetic valorisation).

In those works and during the debate, some aspects of the legislation in force in Portugal, within the scope of waste management, were addressed, namely as regards authorisation of management operations and the lack of legal framework for the study of the environmental impact of the application of waste as materials for civil construction and public works. Recently, there has been an increasing number of authorised units and of authorised waste development operations, which indicates a higher awareness on the part of economic agents as regards this issue. Nevertheless, the legislation in force is still one of the main obstacles to the use of waste in geotechnical works. The authorities are making every effort to facilitate the processes, but in such a way to ensure that the waste, which is likely to be directly used as building material, is neither susceptible to significant physical, chemical or biological transformations, nor likely to have a negative effect on other substances to which they are to enter in contact, so as to ultimately avoid an increase in the environmental pollution or in the hazard to human health.

The preparation of Technical Guides of different industrial sectors, integrated in the National Plan for Prevention of Industrial Waste, can be a contribution to facilitate the use of waste in geotechnical works. Actually, by identifying the unit operations and the corresponding waste, and by characterising the volumes and their geographical distribution, those documents provide significant support to that task.

Some participants emphasised the significant importance of the existence of waste pockets (Almeida et al., 2004) in minimising the problems associated with environmental management and with economy in production, operating as a guide to promote the transaction of raw material. Actually, the existence of a waste pocket may: a) promote the reduction of waste by making the best use of materials (economic valorisation); b) promote the reduction of costs by using by-products and by obtaining a marginal income; c) extend the range of suppliers; d) act as support to the activities of preservation of the environment; e) encourage the establishment of new industries to develop the use of industrial waste; f) lead to the development of new technologies.

During the seminar, there was the opportunity to become acquainted with the French experience, within the framework of the development of the application of waste in geotechnical works, namely in transportation infrastructures. The studies developed in the scope of the project *OFRIR (Observatoire Français pour le Recyclage dans les Infrastructures Routiers*) and which have made it possible to create the site http://ofrir.lcpc.fr that has been made available since August 2003, are an example of the major effort that has been undertaken by some European institutions and governments to solve waste related problems. The project OFRIR has been developed with the purpose of collecting knowledge and of creating a database regarding products that are likely to be used in road works. The information available is intended to provide all those involved in the projects with a better insight into the characteristics of waste, so as to encourage recycling and the use of local materials in transportation infra-structures. The article presented to the Seminar (Jullien, 2004) describes that project, in overall terms, and it provides an interesting view on the data collected and the elements necessary to characterise waste. It also presents the production cycle that leads to the creation of some waste, and mentions some aspects of a geotechnical, environmental and sanitary character.

Gomes Correia (2004) presented a general insight into some technical aspects that must be considered in the framework of the evaluation of performance of waste in geotechnical works. According to the author, there are, at an international level, several major studies that must be taken into consideration in future actions. The conclusions of this work suggest that: i) there are potentially a number of recycled materials that are considered world-wide as substitutes for natural aggregate materials or materials in the construction of transportation infrastructures, provided that they are economic, meet engineering and environmental specifications, and perform as well as traditional materials in the field; b) priority should be given to performance-related tests such as cyclic load triaxial and gyratory compaction; trial field tests are necessary to calibrate models and laboratory tests, mainly to evaluate long-term performance of materials.

The Spanish experience within the use of waste in geotechnical works was presented by Cano & Celemín (2004). The aim of the paper was to describe the state-of-the-art regarding the use of construction waste in embankments and earthworks, with reference to the waste and by-products that have the highest potential for use in Spain, i.e., colliery spoil, fly ash, air cooled blast-furnace slag, construction and demolition waste and scrap tyres. Though a considerable effort has been carried out in characterising and studying the materials produced in that country, the main conclusion is that Spanish experience, as to the significance of actual applications, is not very large, especially when compared to other European countries. Nevertheless, it is possible to refer to a number of examples of use of the first three abovementioned materials, and, to a limited extent, to waste from construction and demolition works, whereas it seems that there are no cases of scrap tyres being utilised, apart from use in the powdered form as a constituent of bituminous mixtures in pavement layers of experimental road sections. An explanation is given of how the final technical document prepared on the basis of the research work undertaken by the CEDEX – "*Catálogo de Residuos Utilizables en la Construcción*" (Wastes Handbook Usable in Construction) – was drawn up. It provides a description of the waste and by-products that have the highest potential for use in the construction of earthworks.

Pinelo (2004) presented some studies that were carried out in LNEC between 1990 and 2002, with the purpose of characterising different industrial wastes and to assess the possibility of their use in the construction of transportation infrastructures. He referred to the methodology adopted and to some applications on site that were carried out as a result of these studies.

Vieira (2004) presented a report on the development process of the coal fly ash produced at Sines thermal power plant of CPPE (Portuguese Company of Electricity). All the fly ash produced (350 thousand tons per year) is used in the production of concrete and cement. The use of this by-product and its integration in cement and in concrete have made it possible to improve the physical and mechanical characteristics of the latter, and have had positive effects on the economical results of the producing company and on the final price of electric power. However, different is the case for the coarser waste produced in the same power plant (bottom ash and slag). Though the amounts in which these waste are produced are considerably smaller than those of fly ash, applications are needed for these wastes.

Manassero presented a conference related with contamination problems associated to the reuse of wastes in geotechnical works (Manassero & Rabozzi 2004). Classification of the by-product in terms of pollutant release potential, design of containment systems and mechanical behaviour of the materials for structural fills were discussed. The paper deals with a simplified classification system that takes into account the compatibility of the clayey mineral barriers with the organic chemicals that can be present within the reused waste. This method combines basic parameters that can be assessed by simple laboratory tests such as clay activity, concentration, dielectric constant and density of the leachate, and defines a compatibility index able to quantify the potential for adverse leachate-soil interaction. The paper also presents a mathematical model for evaluating the pollutant flux and concentration increase in an aquifer underlying a lined by-product embankment. The main aspect of the mechanical behaviour of a particulate by-product has been investigated in the framework of the elasto-plastic-work hardening models, which allows the verification of the stability and settlements to be expected, within short and long term.

6 SOME APPLICATIONS OF WASTE IN GEOTECHNICAL WORKS IN PORTUGAL

6.1 Use of end-of-life tyres in a road embankment

In Portugal, end-of-life tyres were recently used in a road embankment, with the aim of minimising the Martson effect, i.e., the differential settlements between the fill built over the concrete structure and the surrounding earth fill block. Seven layers were built using tyres, each with a thickness of 0.50 m. The embankment was built with soils of A-1-B group (AASHTO), with a minimum degree of compaction of 95%, and slightly humid (w_{opn} + 1%). Drains were constructed for maintaining adequate drainage of the layers of tyres. The installation of inclinometers, piezometers, settlement plates and surface marks was envisaged. For the layers built with tyres, an equivalent modulus of 20% of the soil modulus placed in the embankment was considered. To date it has not been possible to obtain the results of the observation equipment.

6.2 Municipal solid waste incinerator slag in road construction

In Portugal, until mid-90s, solid urban waste was deposited in sanitary landfills, which present a limited operation life and high construction and operation costs. Paying attention to the continuous increase in production of solid urban waste, and as an alternative to landfill deposition, incineration enables to profit from this waste in terms of energy and allows a reduction in the space required for landfilling. The municipal solid waste incineration process results approximately in the production of 20% of slag.

Slags from the Valorsul plant, resulting from the incineration of municipal solid waste originated in the Main Lisbon Area, amount to a volume of about 80 000 t/y. With the aim of studying the possible use of incinerator slag in road construction, some studies were carried out to characterise it, namely chemical characterisation, in order to ascertain its potential for pollution, and physical characterisation (LNEC, 2001a; Sousa et al., 2004). The results show that the materials are non-plastic and have a sand equivalent of about 60%. Methylene blue tests resulted in values of about 0.25 ml/g/100 gr of fines. The Los Angeles tests resulted in average values of 42% (B grading). The CBR value varies between 78 and 97%. The slag showed signs of particle size changes caused by compaction, leading the authors of the study to suggest that it should be verified whether the compaction process used in construction works produces this effect to a significant degree. The results envisage the possible use of this slag in road construction: a) chemical characteristics are, in general, in compliance with the environmental specifications adopted, for instance, in France; b) physical characteristics of the slag are not different from those of a natural aggregate, and would fit into the specifications for materials used in some landfill infrastructure components or in sub-base layers of pavements. Despite the major importance of this study, a deeper mechanical characterisation of the slag is still needed, particularly in order to estimate values of the resilient modulus and to evaluate the long-term behaviour of the material. In order to proceed with the studies, trial sections (sub-base and embankment layers) are being carried out, with the purpose of assessing, on the one hand, the applicability of slag to conventional equipment and to construction methods, and on the other hand, the performance of these materials, by means of the observation of the behaviour of pavements.

In another study, fresh and aged slag resulting from the LIPOR municipal solid waste incinerator (Porto city), were characterised from a geotechnical and environmental point of view (Almeida et al., 2004). From the laboratory tests results, it was concluded that slag could be successfully used in embankment and in subgrade layers. Slag mixed with cement showed a mechanical performance that would make it possible to use it on base and sub-base courses.

6.3 Water treatment plant sludge used in geotechnical application

It has been acknowledged for some time that depositing sludge in an uncontrolled manner may be harmful to the environment, particularly as regards the deterioration of the landscape and air quality, and the pollution of surface and underground water.

Because of the large amounts of sludge produced in water treatment plants in Portugal (20 000 t/y) and the legal restrictions on their landfilling, the final elimination of this sludge represents a major environmental and economic problem for the sector. Many different studies have therefore been developed to characterise the sludge, with the aim of evaluating the possibility of using it in geotechnical works (LNEC, 2003; Carvalho & Roque, 2004).

The chemical tests on sludge have not been conclusive, because although the results are not in compliance with all the requirements defined by the Portuguese law, it can neither be stated that there is no contamination hazard with its application in geotechnical works, nor it can be said that it represents a major pollution source for soils and for surface and ground waters. The results of the physical and mechanical characterisation revealed that this material has a high content of organic matter, a high void ratio, and a low dry unit weight (Proctor and modified Proctor). Some index compaction tests results are as follows: plasticity index values of 26%; standard Proctor maximum dry unit weight of 6.3 kN/m^3 and optimal water content of 84%. The low dry unit weight of the sludge (approximately half of that found in natural soils) could be an advantage, for example when used in embankments on soft soils. When compacted in standard Proctor, it is a relatively low compressibility material ($a_v = 2 \times 10^{-4} m^2/kN$) and its shear strength is considerable ($\phi' = 44°$; $c' = 77$ kPa).

By comparing the results of the chemical and geotechnical characterisation of the sludge with the technical specifications in force in Portugal for materials to be used in pavement layers and transportation earthworks, it could be judged that the sludge is not an appropriate material. However, it is important to consider that these specifications were developed for conventional materials. The origin of these materials is very different from the process of sludge production. It is thus necessary to develop mechanical characterisation studies in order to assess the behaviour of the sludge under conditions representative of the field situation, for instance, when exposed to repeated loads. The use of this material in transportation earthworks, possibly in mixtures with other materials or additives, will have to be more deeply analysed.

6.4 Characterisation of thermal power plant slag and ashes for application in road pavement

Waste resulting from the combustion of coal at a thermal power plant has characteristics which depend on the type of coal and on the method of the combustion and collection equipment. Some characterisation studies on the slag and ashes produced at the power plant of the CPPE (Portuguese Company of Electricity) have been developed in Portugal, with a view to the application of these materials in road constructions. A study was developed which considered the following: i) Generic characterisation of slag and ashes produced at the power plant, aiming at the application in sub-base layers, but exploring also the possibility of use in base layers; ii) Chemical and physical characterisation of the slag and ashes, mechanical characterisation of the slag and construction of a field trial; iii) Structural characterisation of the pavement on the field trial and observation of the behaviour during exploitation. Following an initial characterisation of the slag and ashes in a laboratory (LNEC, 1996), a more complete mechanical characterisation was developed, during which not only was the slag characterised, but also the traditional crushed aggregate, and a mixture of 50% of each material (50/50). To this end, cyclic triaxial load laboratory tests were carried out on cylindrical test specimens with a 150 mm diameter (LNEC, 1999). It should be noted that the initial deformation of the traditional aggregate was somewhat greater than that of the other materials. For confining stress values (σ_3) of 35 and 80 kPa, the values of the reversible deformability modulus were 140 and 300 MPa, respectively, in the case of the slag; 300 and 520 MPa, in the case of the 50/50 mixture; and 250 and 515 MPa in the case of the traditional crushed aggregate. It was concluded that, in particular, the 50/50 mixture produced results similar to those observed for the traditional aggregate. The results of the tests performed on the slag were relatively lower.

A field trial section was then built (LNEC, 2001b) along a length of 230 m, divided into three sub-sections, each one consisting of the following elements: 1) base and sub-base layers with slag; 2) base with a mixture of slag and traditional crushed aggregate (50% + 50%) and a sub-base with slag; 3) base with traditional crushed material and a sub-base with slag. Both layers were 0.20 m thick, and an asphalt surface treatment was applied on the base surface. The differences between the deflections measured in the different sections were not, on average, significant (LNEC, 1996). The main conclusions, which could be drawn from the studies, are as follows: i) the results of the chemical characterisation of the slag and ashes indicate that the use in road construction does not lead to environmental risks; ii) the mechanical behaviour of the materials tested was fairly satisfactory, when compared with natural aggregates; iii) although it does not comply with some of the requirements generally demanded for materials of base and sub-base layers, the use of this slag in these layers seems to be feasible; iv) the use of slag in road construction does not require a specific construction process, therefore enabling traditional construction processes to be used; v) the behaviour of the field trial in which slag was used in sub-base and base layers, is deemed satisfactory, particularly in the section in which a mixture of equal parts of slag and crushed aggregate was used in the base layer.

6.5 Waste gravel and crushed rocks resulting from quarrying

Wastes produced by the natural stone industry (quarrying and transformation of stone) are not hazardous. Nevertheless, the type of treatment leads to the production of fairly high amounts of waste, particularly solids deriving from quarrying and mud driving from transformation, which have significant effects on the landscape and cause logistic and economic problems for the companies. This sector, which can be divided into the sub-sector of ornamental rock and into the sub-sector of industrial rock, has a significant economic importance for the country – about 100 million tons of stones are processed each year –, both in terms of internal and external markets. The main recycled rocks in Portugal identified by the Technical Guide for the Natural Stone Sector are granite, limestone and marble. Even though the production of a large volume of waste is specific of such type of activity, it has been worsened in the last few years, because, traditionally, the quarrying and transformation procedures were carried out only on the basis of economic criteria, without taking into account possible environmental costs, lacking planning and control of operations, with inappropriate technology and unskilled labour. Consequently, these activities have created rather delicate situations in terms of waste management.

Even though quarrying waste can be used in some industrial activities (paint and varnish industry, cement, paper, ceramic industry, etc.) and in farming activities, it is mainly used in civil construction and public works, such as dams and roads (embankments and pavement layers).

For construction of the A6 Motorway in Southeast Portugal waste was used from the exploitation of marble from three different quarries, in a granular layer of 0.125 m placed over the subgrade. A layer of selected limestone aggregate was placed over this layer, also with a thickness of 0.125 m. Physical and mechanical characterisation of this quarry waste was performed in laboratory (LNEC, 1998). Cyclic triaxial load tests were made on cylindrical test specimens with a 300 mm diameter. The values obtained are similar to those usually obtained in selected aggregates used in pavement construction. The mechanical characterisation of the layers was carried out using falling weight deflectometer tests. The values of the modulus for the quarry waste layers were similar to those obtained for the aggregate, i.e., 200 to 300 MPa.

6.6 Quarry mud

Quarries for production of aggregates originate also considerable quantities of fine-grained waste (quarry mud). The possibility of using this mud in geotechnical works, namely mixed with other materials or with additives, has started to be analysed in Portugal. A study is planned on the subject, within a collaboration project between the National Laboratory of Civil Engineering and the owner of several quarries. It is expected that the study include laboratory characterisation of the material, as well as of mixtures of the waste with aggregates and different additives. Trial embankment sections will be also constructed and monitored, and the results of the experimental work will be analysed in order to draw conclusions about the possibilities of reuse in geotechnical works, and, eventually, to draw up technical guidelines of use.

6.7 Construction and demolition waste

In Portugal, a few physical and mechanical characterisation studies have been developed in laboratory and "in situ" on waste materials resulting from demolition and construction activities, and some other studies are currently under way on the subject.

Pereira presented the results of a study intended to assess the construction and demolition waste in the northern zone of Portugal (Pereira & Jalali, 2004). The study observed that, up to recently, the civil construction sector, in general, assigns little importance to issues related with the waste produced. As a result, the available data about the volume produced, as well as about the classification and the destination of waste is scarce.

In another study by Costa (Costa, 2002), it is concluded that, after compacted, demolition materials have shown fairly high values of the modulus of deformability, having, therefore, adequate mechanical characteristics for construction of fills.

Some applications of construction and demolition wastes have been, anyhow, carried out in the country. Mention has to be made to the case of the construction works for the World Exhibition of 1998, EXPO 98, in the area of Lisbon. A high capacity recycling system was installed to process the materials resulting from the demolition of the old buildings that existed at the area. About 812 000 tons of concrete, 190 000 tons of masonry materials (bricks, stones), plus 60 000 tons of recycled material from concrete and bituminous pavements were involved in this recycling operation. As for steel, 5 000 tons were recovered and sent to external steelmaking plants.

After the EXPO 98 some other cases of reusing demolition wastes in Portugal have been undertaken, and it can be said that the future development of this activity seems to be thriving.

6.8 Steel slag

There are two main steelmaking plants in Portugal, which use the electric arc furnace process. The national production of steel slag is 260 000 tones by year. There is an increased pressure to use these materials, particularly in pavements, in spite of having been formerly rejected, as unsuitable, by the routine tests used for natural materials. Promoting the use of steel slag materials in transportation infrastructures and geotechnical constructions can offer engineering, economic, and environmental benefits. However, as mentioned above, national standards are mainly oriented for natural materials. There is still concern that many engineering test methods used for natural materials do not predict the true field performance of these materials and therefore it is necessary to demonstrate that non-natural materials, such as steel slag, will assure, at least, the same quality construction and long term performance. For this reason a research project has been recently initiated in Portugal giving priority to laboratory performance-related tests for structural engineering properties. It also attends to environmental properties, such as pollutant leaching, which are relevant for this type of non natural materials. Three research institutions participate in this Project, which will be finished in 2007: the National Laboratory of Civil Engineering (LNEC), the University of Minho (UM) and the Centre for Reusing Waste of the University of Minho. Financing is partially provided by the *Fundação para a Ciência e a Tecnologia* (FCT). Field calibration of the laboratory tests will be undertaken by constructing trial road sections with several cross-sections, involving mechanical and environmental monitoring performances. These laboratory and "in-situ" test results are intended to be used to establish guidelines to promote the use of steel slag in transportation infrastructures and geotechnical works.

7 FINAL CONSIDERATIONS

The reuse of recycled materials in geotechnical works has awakened a growing interest in the last few years in Portugal. The activities of the TC-3 of the ISSMGE on this subject have contributed up to a point to its development in this country. Actually, integrated in the framework of the TC-3 activities, a committee within the Portuguese Geotechnical Society (SPG) was created in 2003 for analysing the current state of the reuse of recycled materials in geotechnical applications in Portugal and for promoting it nationwide. The results of the activities of this committee, namely the national survey and the seminar, were briefly presented.

In Portugal, there has been some effort in studying the possibility of using recycled materials in geotechnical works. The difficulties usually encountered in implementing a policy of promoting the use of recycled materials were referred in this work, and mention was made to the main requirements of any programme for developing this kind of use, as well as to the ways to overcome the difficulties and to the main agents that must be involved.

A few studies and recent applications developed in Portugal in the field of recycled materials in pavements and transportation earthworks were presented. Several studies have been published which provide evidence for the possibility of using recycled materials in pavements and transportation earthworks, with satisfactory results. Some applications have been carried out for certain wastes, as used tyres, whereas for other wastes only trial sections have been constructed within the scope of research studies. On the other hand, for some given wastes, like construction and demolition waste, the reuse activity in civil engineering applications, including earthworks, is being more active and presents a promising future.

The reuse of recycled materials in geotechnical works could become a fast developing activity in Portugal, in spite of the delay in comparison with other countries, provided that the different agents involved will be aware of the existing difficulties and of the ways to surpass them.

ACKNOWLEDGEMENTS

The financial support of project "*POCI/ECM/56952/2004-Application of waste in transportation infrastructures and geotechnical constructions – Re-use of steel slags*" from *Fundação para a Ciência e a Tecnologia* is acknowledged. The authors also wish to thank Ms. Maria Lurdes Antunes (PhD) and Mr. António Roque (PhD) for the elements provided.

REFERENCES

Almeida, A.; Lopes, M.L.; Bastos, A.M.; Figueiras, J. 2004. LIPOR's approach to the re-use of waste. Seminar about the use of recycled materials in geotechnical works. Guimarães (in Portuguese).

Cano, H. & Celemín, M. 2004. Applicability study for using wastes and by-products in earthworks and embankments. Seminar about the use of recycled materials in geotechnical works. Guimarães.

Carvalho, M. & Roque, A.J. 2004. Re-use of sludge of water treatment plants. Prospects for application in geotechnical works. Seminar about the use of recycled materials in geotechnical works. Guimarães, (in Portuguese).

Costa, H.M. 2002. Demolitions and construction wastes applied in geotechnical works. MCS thesis presented to the Faculty of Sciences and Technology. Universidade Nova, Lisbon, (in Portuguese).

Gomes Correia, A. (2004). The use of processed materials in transportation. An International overview. Seminar about the use of recycled materials in geotechnical works. Guimarães.

Jullien, A. 2004. Panorama du recyclage dans les infrastructures routières françaises: le projet OFRIR. Seminar about the use of recycled materials in geotechnical works. Guimarães (in French).

LNEC 1996. Characterization of coal power plant slag and ash for re-use in road pavement layers. Report 251/96 – NPR, (in Portuguese).

LNEC 1998. Study on the mechanical behaviour of Highway A6 pavement granular layers, section Évora-Estremoz. 2nd Report. Report 164/98 – NPR, (in Portuguese).

LNEC 1999. Use of slags produced in Sines coal power plant in a pavement trial section. 1st Report. Laboratory characterization. Report 143/99 – NPR, (in Portuguese).

LNEC 2001a. Study on the feasibility of re-use of municipal solid waste incineration slags in roadway construction. Report 299/01, (in Portuguese).

LNEC 2001b. Use of slags produced in Sines coal power plant in a pavement trial section. 2nd Report – Observation of construction works. Report 170/01 – NPR, (in Portuguese).

LNEC 2003. Recycling of water treatment plant sludges. Geotechnical and geo-environmental characterization of Asseiceira ETA's sludge (Tomar). Final report. Report 111/03 – NGE, (in Portuguese).

Lobo, F. & Santiago, A. (2004). Legal regime for re-using wastes in Portugal. Seminar about the use of recycled materials in geotechnical works. Guimarães (in Portuguese).

Manassero, M. & Rabozzi, C. 2004. Contamination problems related to the reuse of wastes in geotechnical works. Seminar about the use of recycled materials in geotechnical works. Guimarães.

Oliveira, R. 2004. Integrated management of wastes: the environmental impact. Seminar about the use of recycled materials in geotechnical works. Guimarães (in Portuguese).

Pereira, L. & Jalali, S. 2004. Evaluation of construction and demolition wastes in Northern Portugal. Seminar about the use of recycled materials in geotechnical works. Guimarães (in Portuguese).

Pinelo, A. 2004. Use of recycled materials in transportation routes. Seminar about the use of recycled materials in geotechnical works. Guimarães (in Portuguese).

Sousa, C.D.; Freire, A.C.; Antunes, M.L. 2004. Slags, an alternative resource in highway construction. Seminar about the use of recycled materials in geotechnical works. Guimarães (in Portuguese).

Vieira, F.C. 2004. Reuse of coal ashes. Seminar about the use of recycled materials in geotechnical works. Guimarães (in Portuguese).

13

Utilization of Processed Material in Pavement Surface and Base-course in Japan

Nobuyuki Yoshida
Research Center for Urban Safety and Security, Kobe University, Kobe, Hyogo, Japan

ABSTRACT: This paper describes the current state of utilization of processed material in pavement construction in Japan. The designing method of asphalt pavement currently used in Japan is first briefly explained together with the process of adopting non-standard material in pavement construction. Regarding engineering and environmental aspects, enumerated are the laws, guidelines, *etc.* required to follow in case of adopting non-standard material. A recent movement toward a performance-specified pavement construction is also mentioned which has been motivated by the issue of "Technical standards regarding pavement structure" in 2001 by Ministry of Land, Infrastructure and Transport and of which cases are being gradually accumulated.

1 INTRODUCTION

In Japan, about 406 million tons of industrial wastes and about 52 million tons of general wastes were produced in 2000, and only approximately 45% of the former and 14% of the latter were recycled, respectively. For industrial wastes, sludge account for about 47%, animal excreta for about 22%, construction wastes for about 16%, slag for about 4%, and so on. Promoting effective measures such as reduce, reuse, recycling, *etc.* particularly against industrial wastes is an urgent matter since the remaining capacity of disposal sites for industrial wastes is getting smaller.

In recent years, yearly use of aggregates amounts to approximately 730 million tons. It is about 78% of its peak but production of good quality natural aggregates is diminishing due to conservation of natural environment and saving of natural resources. Thus, it is urgent to promote recycled or processed material to be utilized wherever possible.

In 2000, the Japanese Government enacted the Basic Law for Establishing the Recycling-based Society, and accordingly some laws have been revised or newly enacted. Those are Waste Management and Public Cleaning Law (revised), Law for Promotion of Effective Utilization of Resources (revised), Container and Packaging

Recycling Law (existing), Electric Household Appliance Recycling Law (existing), Construction Material Recycling Act (newly enacted), Food Recycling Law (newly enacted) and Law on Promoting Green Purchasing (newly enacted). Construction Material Recycling Act and Law on Promoting Green Purchasing have a significant influence on construction industry, and procurement of processed or recycled products is being increased in public works year by year.

Moreover, laws promoting use of recycled material would not necessarily ensure that all processed materials are approved in actual use without other considerations. Constraints from environmental and also engineering aspects are imposed. Most environmental concerns are designated in Water Pollution Control Law (revised in 2001a) and Soil Pollution Measures Law (enacted in 2002).

In the followings, the designing method of asphalt pavement adopted in Japan is first outlined together with the process of adopting by-products and new material in pavement construction.

Recent movement towards performance-specified construction is also briefly described, and engineering and environmental requirements are presented in relation to some of presently used processed materials.

2 DESIGNING METHOD OF ASPHALT PAVEMENT IN JAPAN

In Japan, pavement design has been carried out based upon a subgrade CBR value and an equivalency conversion thickness (T_A), so-called CBR-T_A method, since 1967. A multi-layered elasticity based designing method was also conceptually introduced in 1992 but it is not presently used in practice and the CBR-T_A method is a dominant method. Figure 1 shows the CBR-T_A design flow used in Japan (Japan Road Association, 1992).

T_A is an equivalency conversion thickness, representing the pavement thickness required if the total pavement depth was to be constructed with hot asphalt mixtures used for binder and surface courses. It is estimated by

$$T_A = 3.84 \frac{N^{0.16}}{CBR^{0.3}} \tag{1}$$

where N is the total number of 49 kN-equivalent wheel load applications (wheels per one direction) in a specific design period, say 10 years, and CBR is a design subgrade CBR.

A pavement section is designed so as for T_A' to be greater than T_A above. T_A' is computed as

$$T_A' = a_1 T_1 + a_2 T_2 + \cdots + a_n T_n \tag{2}$$

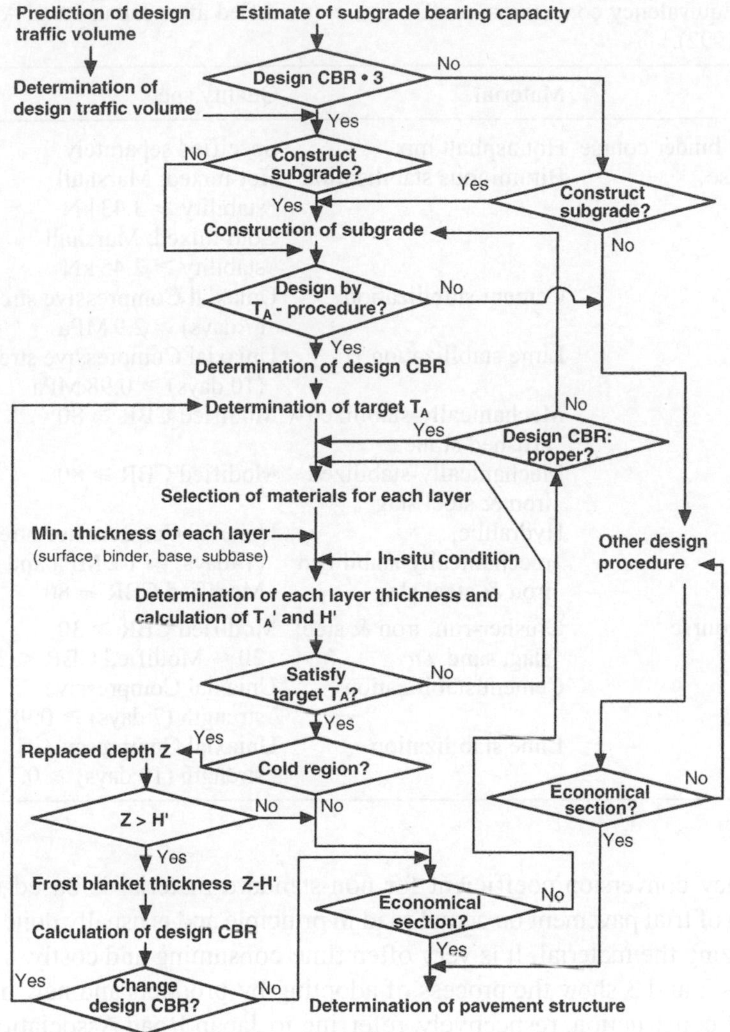

Figure 1. Flow of structural design of asphalt pavement (Japan Road Association, 1992).

where T_A' is the equivalency conversion thickness of the designed pavement section, a_i is an equivalency conversion coefficient for each pavement material used, and T_i is the thickness of each layer. From this, it is seen that, in principle, designing a pavement section requires equivalency conversion coefficients for all the material used. Equivalency conversion coefficients for currently used standard pavement material are summarized in Table 1 (Japan Road Association, 1992).

In the case that non-standard material is attempted to be used in pavement construction, there are two ways to follow: one is to derive an equivalency conversion coefficient for the material and the other is to verify that it possesses the quality equivalent to or better than standard pavement material. Determination of an

Table 1. Equivalency conversion coefficient, a_i (modified after Japan Road Association, 1992).

Location	Material	Quality spec.	a_i
Surface & binder course	Hot asphalt mix	Specified separately	1.00
Base-course	Bituminous stabilization	Hot-mixed: Marshall stability $\geqslant$ 3.43 kN	0.85
		Cold-mixed: Marshall stability $\geqslant$ 2.45 kN	0.55
	Cement stabilization	Uniaxial Compressive strength (7 days) $\geqslant$ 2.9 MPa	0.55
	Lime stabilization	Uniaxial Compressive strength (10 days) $\geqslant$ 0.98 MPa	0.45
	Mechanically-stabilized crushed stone	Modified CBR $\geqslant$ 80	0.35
	Mechanically-stabilized iron & steel slag	Modified CBR $\geqslant$ 80	0.35
	Hydraulic, mechanically-stabilized iron & steel slag	Uniaxial Compressive strength (14days) $\geqslant$ 1.2 MPa and Modified CBR $\geqslant$ 80	0.55
Subbase-course	Crusher-run, iron & steel slag, sand, *etc.*	Modified CBR $\geqslant$ 30	0.25
		20 $\leqslant$ Modified CBR $<$ 30	0.20
	Cement stabilization	Uniaxial Compressive strength (7 days) $\geqslant$ 0.98 MPa	0.25
	Lime stabilization	Uniaxial Compressive strength (10 days) $\geqslant$ 0.7 MPa	0.25

equivalency conversion coefficient for non-standard material is based upon performance of trial pavement on actual road in principle and is usually done in case of standardizing the material. It is very often time-consuming and costly.

Figures 2 and 3 show the process of adopting by-products and new material in pavement construction, respectively, referring to Japan Road Association (1992). In the both cases, the environmental safety of the material of concerned is the utmost concern, and it is also important to ensure that it possesses the quality and performance equivalent to or better than standard pavement material. For this, trial construction is usually carried out. For instance, cracks, rut depth and longitudinal roughness observed are measured on trial pavement with non-standard material at a regular interval and used to compute variation of a present serviceability index with time. It is, then, compared with that of trial pavement with standard material or relevant past data, from which performance equivalency is to be evaluated.

Laboratory test based derivation may also be possible as such that an equivalency conversion coefficient is deduced through comparison of the elastic modulus or uniaxial compressive strength of non-standard material with those of standard one; however, this has been rare to author's knowledge.

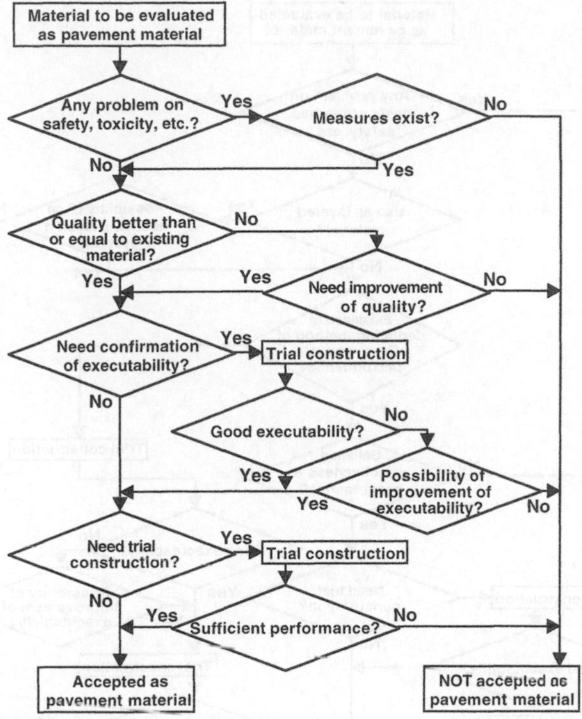

Figure 2. Process of adopting by-products in pavement construction (Japan Road Association, 1992).

In 2001, Manual for Asphalt Pavement was fully revised and renamed as Pavement Design and Construction Guide (Japan Road Association, 2001b), supplemented with Pavement Construction Manual (Japan Road Association, 2001c). A multi-layered elasticity based designing method was illustrated with some examples in a bit better way than in 1992: however, because of ambiguity in damage models, software, *etc.*, little application has been reported so far except research purposes. On the other hand, implementation of performance-specification was probably the most important and influential matter in the revised manual.

3 PERFORMANCE-SPECIFICATION IN ASPHALT PAVEMENT CONSTRUCTION

3.1 Performance indices and examples

In response to the issue of "Technical standards regarding pavement structure" in 2001 by Ministry of Land, Infrastructure and Transport (Japan Road Association, 2001a), performance-specification was implemented in the Pavement Design and Construction Guide as mentioned earlier. The performance indices are specified in

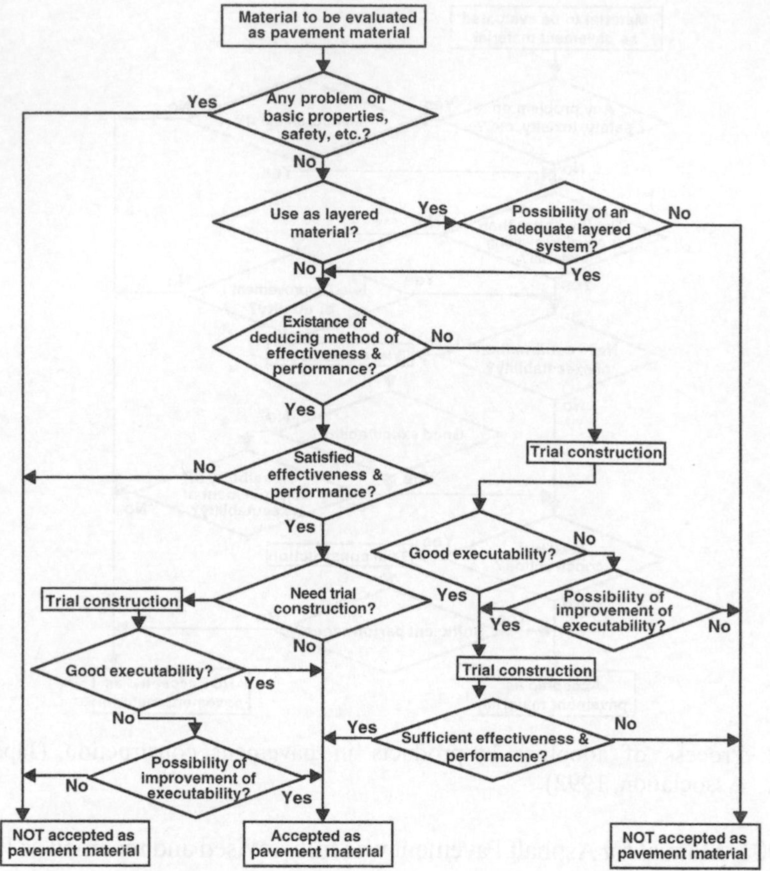

Figure 3. Process of adopting new material in pavement construction (Japan Road Association, 1992).

the technical standards: essential indices for pavements of carriageway and marginal strip are (a) the total number of wheel load applications to produce a crack, (b) the total number of wheel load applications to produce a rut and (c) roughness. For structurally permeable pavement, (d) permeability is also an essential index. Moreover, in addition to these, other indices may be designated from the viewpoints of skid resistance, noise reduction, *etc.*, if necessary.

The standard values for the total number of wheel load applications to produce a crack are summarized in Table 2. Note that a standard wheel load is set as 49 kN, or 5 ton in Japan, and "crack" here denotes the one resulting from fatigue failure, being initiated from the bottom of asphalt surface layer and propagated upwards. The standard values were derived from relationships between heavy traffic volume and the number of 49 kN-converted wheel load applications for a design period of 10 years. For a design period other than 10 years, some adjustment is necessary (Japan Road Association, 2001a, b).

Table 2. Standard values of total number of wheel load applications to produce a crack (modified after Japan Road Association, 2001a, b).

Design traffic volume of heavy vehicles (per day per direction)	Total number of wheel load applications to produce a crack (per 10 years)
more than 3000	35,000,000
Form 1000 to 3000	7,000,000
From 250 to 1000	1,000,000
From 100 to 250	150,000
100 or fewer	30,000

Table 3. Standard values for total number of wheel load applications to produce a rut (modified after Japan Road Association, 2001a, b).

Road classification	Design traffic volume of heavy vehicles (per day per direction)	Total number of wheel load applications to produce a rut (per mm)
1st and 2nd categories, 3rd category's 1st and 2nd classes, and 4th category's 1st class	more than 3,000	3,000
	less than 3,000	1,500
Others	–	500

Table 3 shows the standard values for the total number of wheel load applications to produce a rut. "Rut" here denotes a residual settlement of asphalt surface layer due solely to its non-elastic deformation, excluding deformation due to its abrasion and non-elastic deformation of subsurface layers. The standard values were derived based upon past data of dynamic stability obtained from wheel tracking tests. 500 times/mm is for low volume road referring to lower limits of conventional dense-graded asphalt mixture; 1500 times/mm is for trunk road referring to upper limits of conventional dense-graded asphalt mixture; and 3,000 times/mm is for trunk road with high heavy traffic volume referring to lower limits of modified asphalt mixture (Japan Road Association, 2001a, b). Note that the standard temperature of wheel tracking test here is 60 degrees Celsius.

The roughness immediately after construction is required to be smaller than 2.4 mm. The permeability, expressed in the amount of permeated water per 15 seconds, for structurally permeable pavement should meet the standard values given in Table 4. It is determined by performing an in-situ permeability test. 300 ml/15 sec corresponds to the lower limits for porous straight-asphalt concrete having an air void of about 15% and 1000 ml/sec to the lower limits for porous modified-asphalt concrete having an air void of about 20% (Japan Road Association, 2001a, b).

It has been a year or so since the performance-specified construction was introduced but cases are being accumulated. Table 5 shows some of the performance-specified construction cases for drainage asphalt pavement. Although not given here,

Table 4. Standard values for permeability (modified after Japan Road Association, 2001a, b).

Classification	Amount of permeated water (ml/15 s)
1st and 2nd categories, 3rd category's 1st and 2nd classes, and 4th category's 1st class.	1,000
Others	300

Table 5. Performance-specified constructions for drainage asphalt pavement.

Owner	No. of cases	Inspection	Dynamic stability (times/mm)	Permeability (ml/15 sec)	Roughness (mm)	Noise (dB)
Kanto Branch of M.L.I.T.*	1	Immediately after construction	4000	1000	2.4	89
		1 year later	–	–	–	90
Shikoku Branch of M.L.I.T.	3	Immediately after construction	3000	1000	2.4	89
		1 year later	–	–	–	90
Tohoku Branch of M.L.I.T.	1	Immediately after construction	3000	1000	2.4	89
		1 year later	–	–	–	90
Hyogo Prefecture	1	Immediately after construction	1500	900	2.4	89
		1 year later	–	–	–	90

* M.L.I.T. stands for Ministry of Land, Infrastructure and Transport.

but there are more than 40 cases in which only a noise-related index is specified in such a way that a noise level measured by a "RAC" vehicle should be not more than 89 dB immediately after construction and 90 dB or less after one year. Note that all cases so far constructed with performance-specification are drainage asphalt pavements. In Japan, drainage asphalt pavement has drawn quite attention in recent years and is becoming popular in both highway and general road. Permeable asphalt pavement and water-retention pavement are also being adopted in big cities. In Osaka City, for instance, the magnitude of temperature drop is taken as one of performance-specification items for water-retention asphalt pavement in such a way that difference in highest surface temperatures measured on water-retention pavement and ordinal dense-graded pavement in summer time should be greater than 5 degrees Celsius.

3.2 Measurement and verification of specified performance

Specified performance is verified immediately after construction and may be at a certain period of time after open for traffic, if necessary, depending upon contractual conditions.

Verification methods for some of the performance items are suggested in Japan Road Association (2001a, b). For the total number of wheel load applications to produce a crack, three methods are enumerated: (a) in-situ repeated-loading test with accelerated loading equipment, (b) repeated-loading test on model pavement having the same pavement structure as in-situ, and (c) past experience. The method (a) is not realistic at present since there is no standard testing method nor testing equipment. The method (b) denotes large-scale accelerated loading tests or full-scale trial pavement, the latter being a traditional approach once decided to perform. Currently common is the method (c) which considers that the pavement of concerned will retain the required performance if it has the same pavement structure as an existing other pavement designed with the CBR-T_A method or whose performance was previously verified.

The total number of wheel load applications to produce a rut may be verified by: (a) in-situ repeated-loading test with accelerated loading equipment; (b) repeated-loading test on model pavement having the same pavement structure as in-situ with a temperature of 60 degrees Celsius; (c) wheel tracking test with a temperature of 60 degrees Celsius (Japan Road Association, 1996); or (d) past experience. The methods (c) and (d) are dominant ones in practice.

Roughness is verified based upon its measurement using a three-meter profilometer or a profilometer vehicle.

Permeability is verified by conducting in-situ falling-head permeability test. The permeability, in ml/15 sec, is estimated from measurement of elapsed time for water to flow by 400 ml into pavement in such a way that the permeability is 400 ml divided by elapsed time (in sec) and multiplied by 15 (Japan Road Association, 1996).

Japan Road Association (2001a, b) states encouragingly that any other method may be adopted for verification whenever its effectiveness is demonstrated.

4 PROCESSED MATERIAL USED IN PAVEMENT CONSTRUCTION

4.1 Present state

Here only discussed are aggregates possibly used in the surface and base-course for road construction. Table 6 shows some of the processed materials possible for road construction. Some are used as standard pavement material and others not.

Iron and steel slag has been standard pavement material designated in Japanese Industrial Standard with several revisions since 1979 (Japanese Standards Association, 1992). Table 7 summarizes quality requirements along with those for mechanically-stabilized crushed stone for comparison. Although not given on the table, the range of grading is also specified. Not mention but all pavement material must satisfy the effluent standards for pollutants as given in Table 8 designated by Ministry of Environment (2001b).

Table 6. Processed material in road construction.

Material	Production in thousand tons (year)	Utilisation in thousand tons (year)
Air-cooled BF slag	5,884 (2002)	3,773 partially including railway use
Granulated BF slag	18,321 (2002)	106 partially including railway use
Steel slag	12,168 (2002)	3,306 partially including railway use
Crushed concrete	35,000 (2000)	No data but approx. 96% reused/recycled in construction work
Crushed asphalt	30,000 (2000)	No data but approx. 98% concrete reused/recycled in construction work
Coal fly ash	8,429 (2000)	4,893 in cement production and 96 in road construction
Sewage sludge	1,977 in dried cond. (2001)	0.60 in asphalt additive; 0.28 in asphalt filler; 1.94 in surface & base layers
Harbour dredged	8,000 including construction sludge (2000)	No data but approx. 41% recycled in construction work
Rock wool/slag wool	350 (2000)	No data on use in road construction
Glass bottle	1,820 (2000)	No data on use in road construction but approx. 1,416 are crushed and grading-adjusted and used in various fields
Tires	972 (1999)	No data on use in road construction

Quality requirements for processed aggregates derived from crushed concrete and asphalt for use in base-course are specified in Plant-Recycling Pavement Technology Guide (Japan Road Association, 1992). For instance, in case of using such processed aggregates in base-course of asphalt pavement as mechanically stabilized crushed stone, the modified CBR should be not smaller than 80% and the plasticity index not greater than 4. The other aspects follow those for mechanically stabilized crushed stone composed of natural aggregates.

Aggregates derived from melt-solidified products of general waste, sewage sludge, etc. are presently in stage of tentative standard of JIS (Japanese Standards Association, 2002). For aggregates used for hot asphalt mix, the density in oven-dried condition should be 2.45 g/cm^3 or greater, the water absorption percentage 3.0% or smaller and the abrasion 30% or smaller. For aggregates used for base-course, on the other hand, the unit weight should be 1.5 kg/litre or larger, the abrasion 50% or smaller, and so on. The effluent standards must also be satisfied as shown in Table 8.

Use of glass bottle as aggregates in asphalt mixture has been investigated by Public Works Research Institute (1999). For this, glass bottle is crushed and grading-adjusted, which is called "glass cullet". Environmental constraints are the same for other material and the effluent standards for pollutants should be satisfied. For the engineering aspects, typical characteristics of glass cullet are as follows: the specific gravity ranges from 2.45 to 2.55; the water absorption is up to 0.3%; the

Table 7. Iron and steel slag for road construction (modified after Japanese Standards Association, 1992).

Items	Hydraulic mechanically stabilized iron and steel slag	Mechanically stabilized iron and steel slag	Crusher-run iron and steel slag	Single-graded steel slag	Crusher-run steel slag	Mechanically stabilized crushed stone
Designations	HMS-25	MS-25	CS-40, CS-30, CS-20	SS-20, SS-5	CSS-30, CSS-20	M-40, M-30, M-25
Usage	Base-course	Base-course	Subbase-course	Hot asphalt mix	Hot asphalt-stabilization	Base-course
Coloration*1	No colora.	No colora.	No colora.	–	–	–
Immersion expansion ratio*2	1.5 or smaller	1.5 or smaller	1.5 or smaller	2.0 or smaller	2.0 or smaller	–
Unit weight (kg/lietr)	1.5 or larger	1.5 or more	–	–	–	–
Uniaxial compressive (13-day cured) strength (MPa)	1.2 or larger	–	–	–	–	–
Modified CBR (%)	80 or larger	80 or larger	30 or larger	–	–	80 or larger
Specific gravity*3	–	–	–	2.45 or greater	–	–
Water absorption percentage (%)	–	–	–	3.0 or smaller	–	–
Abrasion (%)	–	–	–	30 or smaller	50 or smaller	–
Aging*4	6 months or more	6 months or more	6 months or more	3 months or more	3 months or more	–
Plasticity index (%)	–	–	–	–	–	4 or smaller

*1 Only for blast-furnace slag. *2 Only for steel slag. *3 In saturated surface-dry condition. *4 For steel slag.

Table 8. Effluent standards against soil pollution (Ministry of Environment, 2001).

Toxic substances	Permissible limits
Cadmium and its compounds	0.01 mg/l
Cyanide compounds	Not detectable
Organic phosphorus compounds	Not detectable
Lead and its compounds	0.01 mg/l
Sexivalent chrome compounds	0.05 mg/l
Arsenic and its compounds	0.01 mg/l
Total mercury	0.0005 mg/l
Alkyl mercury compounds	Not detectable
PCBs	Not detectable
Trichloroethylene	0.03 mg/l
Tetrachloroethylene	0.01 mg/l
Dichlorornethane	0.02 mg/l
Carbon tetrachloride	0.002 mg/l
l, 2-dichloro ethane	0.004 mg/l
1, 1-dichloro ethlene	0.02 mg/l
cis-1, 2-dichloro ethylene	0.04 mg/l
1, l, 1-trichloro ethane	1 mg/l
1, 1, 2-trichloro ethane	0.006 mg/l
l, 3-dichloropropene	0.002 mg/l
Thiram	0.006 mg/l
Simazine	0.003 mg/l
Thiobencarb	0.02 mg/l
Benzene	0.01 mg/l
Selenium and its compounds	0.01 mg/l

abrasion ranges from 40 to 50%; and grains smaller than 20 mm are dominant. Asphalt mixtures have been formed with several different mixing proportions of glass cullet, and trial pavement has been carried out. It is suggested that a mixing proportion of 10% is an upper limit from the view of decrease in residual stability or resistance against stripping, and mixing of glass cullet more than 10% may fail to satisfy the requirements for asphalt mixtures designated in Pavement Design and Construction Guide. It is still under investigation.

4.2 Example

Use of fly ash as base-course aggregates has been studied by Slag Base-Course Research Group managed by the road construction division of Hyogo Prefecture. Most fly ash comes from thermal power plants in Japan and the amount is significantly increasing.

For the purpose of using it as an aggregate for base-course of asphalt pavement, a fly ash pellet was developed in such a way that a fly ash was first solidified with cement and gypsum, then crushed and graded for use. It is presently called "ash-stone". A compound slag (hereafter called HMS-25fa) was then formed by mixing steel-making slag, blast-furnace slag and ash-stone at pre-specified proportion. Investigation on the basic characteristics of the compound slag including physical tests, durability test, etc. was carried out in 1998. Deformational characterization was also undertaken subsequently (Yoshida et al., 1998; Yoshida and Nishi, 2000). After confirming the quality and safety of HMS-25fa, then, trial pavement was constructed in two prefectural roads, about 236 m in Ono-Shikata line at the end of 1998 and about 132 m in Obe-Akashi line at the end of 1999. Trial pavement with HMS-25, a standard base-course material, was also constructed at the same locations for comparison. Performance of the trial pavements at the both locations was investigated for about three years: at Ono-Shikata line from the end of 1998 to the end of 2001 and at Obe-Akashi line from the end of 1999 to the end of 2002. Investigation consisted of FWD test and measurement of surface cracks, rutting, roughness, pavement temperature, etc. at regular intervals. Not mention but traffic volumes at the both locations were investigated before construction of trial pavement and several times after construction. From these data, then, a present serviceability index was computed and plotted against elapsed time after construction. In spring 2003, detailed inspection on these data concluded that HMS-25fa had demonstrated durability and performance equivalent to HMS-25. In July 2003, HMS-25fa was implemented in the pavement manual of Hyogo Prefecture.

Since it is verified that the compound slag possesses the equivalent performance to HMS-25, an equivalency conversion coefficient of 0.55 can be used in the structural design, at least in Hyogo Prefecture.

5 CONCLUSIONS

This paper briefly introduced the present state of utilization of processed material in pavement construction in Japan. The current designing method is outlined together with the adopting process of non-standard material in pavement construction. Performance-specified construction, which is becoming a recent trend, is also mentioned with examples.

Empiricism underlies the current designing method and is not suitable for adopting non-standard material in pavement construction in a rational manner. As suggested in the revised Manual, establishing a mechanistic, theoretical designing method appears to be essential for this.

Performance-specified pavement construction will soon be a major contractual trend in Japan, at least for asphalt pavements with additional functions such as drainage pavement, permeable pavement, *etc*. To make this practice effective and rational, accumulation of performance data is vital, and this, in turn, shall lead to reviewing appropriateness of the performance items specified at present.

Development of rational verification technique, not trial pavement, for each performance index is also desirable.

ACKNOWLEDGEMENT

The statistical data on processed material are quoted from Japan Slag Association, Council for Promotion of Public Relations for Recycling of Construction By-products, Center for Coal Utilization, Japan, Council for Promotion of Recycling of Glass Bottles, Japan Sewage Works Association, and Rock Wool Association. The author hereby expresses his gratitude to them.

REFERENCES

Japan Road Association, Manual for Asphalt Pavement, 1992 Edition, 324pp. (1992) (in Japanese)

Japan Road Association, Plant-Recycling Pavement Technology Guide, 132pp. (1992) (in Japanese)

Japan Road Association, Pavement Testing Manual, Supplement Volume, 317pp. (1996) (in Japanese)

Japan Road Association, Technical Standards regarding Pavement Structure and its Commentary, 91pp. (2001a) (in Japanese)

Japan Road Association, Pavement Design and Construction Guide, 333pp. (2001b) (in Japanese)

Japan Road Association, Pavement Construction Manual, 314pp. (2001c) (in Japanese)

Japanese Standards Association, Iron and steel slag for road construction, JIS A 5015, 20pp. (1992)

Japanese Standards Association, General waste and sewage sludge etc. melt-solidified products derived aggregates for road construction, TR A 0017, 20pp. (2002) (in Japanese)

Ministry of Environment, Water Pollution Control Law, (2001a) (in Japanese)

Ministry of Environment, Environmental Quality Standards for Soil Pollution, (2001b) (in Japanese)

Ministry of Environment, Soil Pollution Measures Law, (2002) (in Japanese)

Public Works Research Institute, Manual for test and evaluation of recycled materials from other industries for use in trial construction in public works – tentative, (1999) (in Japanese)

Yoshida, N., Sano, M., Hirotsu, E., Nishi, M., Arai, T. and Toyama, S., Performance and equivalency coefficient of compound slag containing solidified fly ash, Journal of Pavement Engineering, Japan Society of Civil Engineers, Vol. 3, pp. 147–156 (1998) (in Japanese)

Yoshida, N. and Nishi, M., Deformation characteristics of slag base-course material containing fly ash pellet, Journal of Materials, Concrete Structures and Pavements, Japan Society of Civil Engineers, No.641/V-46, pp. 269–275 (2000)

14

Geotechnical Characteristics of Crushed Concrete for Mechanical Stabilization

E. Sekine & Y. Momoya
Railway Technical Research Institute, Tokyo, Japan

ABSTRACT: Crushed concretes produced at construction sites are increasingly used as a roadbed material for the construction of roads. To apply crushed concretes to railway earth structures, it is necessary to confirm their deformation characteristics in detail. However, the deformation characteristics of crushed concretes with the objective of geotechnical engineering are not yet investigated sufficiently. To investigate their strength and deformation characteristics, compaction tests and triaxial tests were carried out. Crushed concretes were collected from processing plants all over the country. The results showed that the strength and deformation characteristics were correlated with water absorption of the material. Through the investigation, the effect of water absorption on particle crushing and strength became apparent.

1 INTRODUCTION

Recently, a large amount of crushed concretes are generated at construction sites. It is necessary to recycle those crushed concretes for the purpose of environmental protection and reduction of construction cost. Besides, it becomes difficult to obtain high quality soil materials for new construction in our country. Under the circumstances, crushed concretes are increasingly used as a roadbed material in the construction of roads. However, it is thought that the quality of crushed concretes is inferior to the quality of natural materials. For instance, strength of particles is smaller than that of natural materials. Details of the strength and deformation characteristics of crushed concrete are not clear yet. This paper discusses the strength and deformation characteristics of crushed concrete materials, which were investigated by compaction tests and triaxial compression tests.

2 PROPERTY OF CRUSHED CONCRETE MATERIAL

Crushed concretes were collected from 26 stone processing plants all over the country. Figure 1 shows the grain curve distribution of these crushed concrete materials.The average of gravel fraction was 75%, which were distributed from

50% to 96%. The sand fraction was less than 40% (average 21%), and fine particle fraction was less than 10% (average 4%). In most crushed concrete materials, the values of uniformity coefficient were less than 40, with some exceeding 100. The average of uniformity coefficient was 28.7. Natural water contents were 2–14% (average 7.4%). Specific gravities of grain particle were 2.5–2.8 (average 2.644).

The water absorption of these crushed concretes were 0–9% (average 5.3%). As the water absorption of natural crushed stones for mechanical stabilization is usually less than 3%, it is assumed that the crushed concretes have more voids than the natural crushed stones. Existence of voids in the particles means inferiority of durability for crushing. To investigate the relationship between the percentage of abrasion and water absorption, an abrasion test was conducted by using a Los Angles Machine. See figure 2 for the test result. The percentage of abrasion increased as water absorption increased. Generally, the water absorption of high

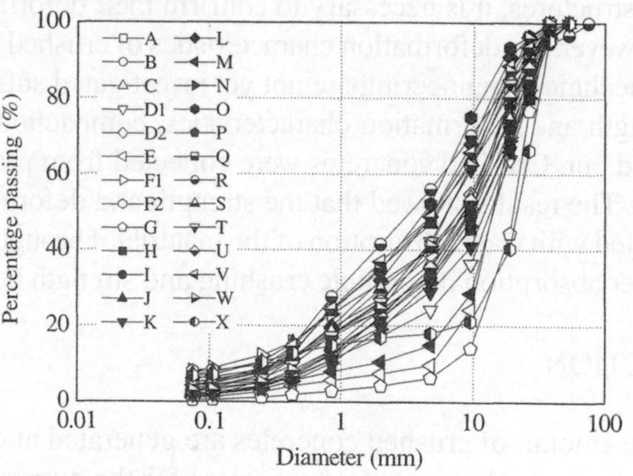

Figure 1. Grain distribution curve of 26 different crushed concrete materials.

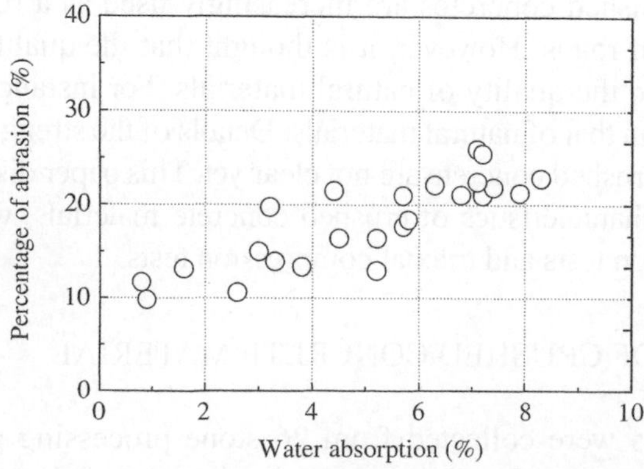

Figure 2. Relationship between percentage of abrasion and water absorption.

quality natural crushed stone is less than 20%. In figure 2, water absorption at 20% abrasion is about 6%. Therefore, it is considered that the crushed concrete with water absorption higher than 6% has insufficient durability against abrasion.

PH tests of crushed concretes were also carried out. The PH of crushed concretes was distributed in a wide range from strong alkali to neutral. Besides, a sexivalent chrome elution test was conducted. In the test, 11 materials out of 26 exceeded the limit value of 0.05 mg/l, which was set up by the Ministry of Environment in 1991. It was proved that the checking of sexivalent chrome is an important issue for using crushed concretes at construction sites.

3 COMPACTION CHARACTERISTICS OF CRUSHED CONCRETE

To apply crushed concrete as an embankment material for railways, it is important to understand compaction characteristics. To investigate compaction characteristics of crushed concretes, compaction tests were carried out at different amount of compaction energy. The Proctor's compaction energy is described by the following equation.

$$E_C = W_R \cdot H \cdot N_B \cdot N_L / V \quad (1)$$

Where

N_B: The number of times for falling.
N_L: The number of compaction layers.
V: Volume of mold.
W_R: Weight of rammer.
H: Height of falling.

Table 1 shows the conditions for compaction tests. Figure 3 shows the relationship between the maximum dry density and compaction energy. The maximum dry density increased as the compaction energy increased. The average of the maximum dry density was 1.674 t/m^3 under the smallest compaction energy E_1, and 1.852 t/m^3 under the largest compaction energy E_4. In order to investigate the crushability of crushed concrete particles in the compaction tests, the relationship between the crushing ratio BM obtained by the Marsal's method and compaction

Table 1. Compaction test conditions.

	E_C (MN m/m^2)	W_R (N)	H (m)	N_L	N_B	V (10^{-3}m^3)
E_1	0.255	24.5	0.30	3	25	2.209
E_2	0.560	24.5	0.30	3	55	2.209
E_3	1.513	44.1	0.45	3	55	2.209
E_4	2.530	44.1	0.45	3	92	2.209

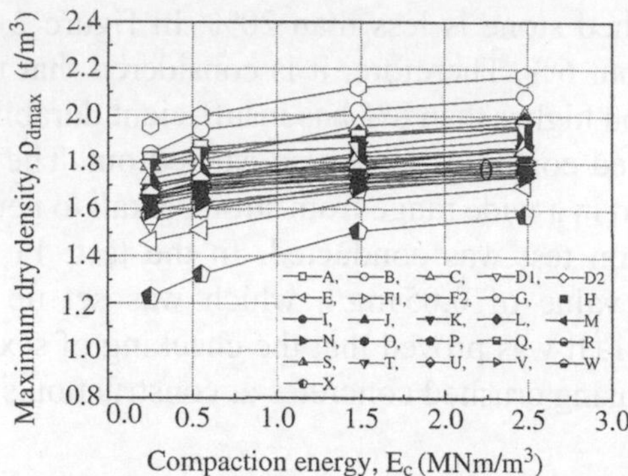

Figure 3. Relationship between maximum dry density and compaction energy.

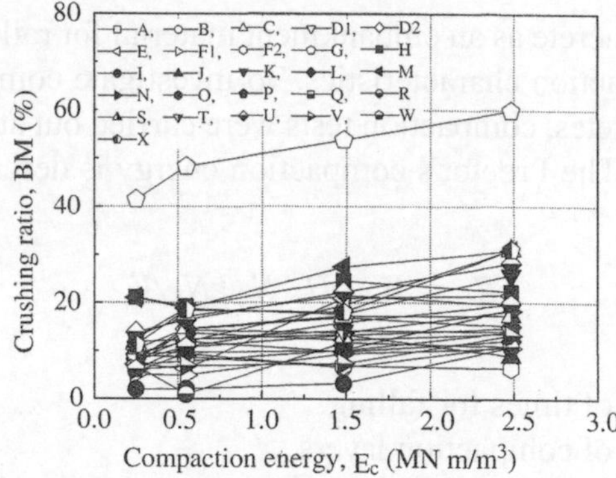

Figure 4. Relationship between crushing ratio BM and compaction energy.

energy was plotted in figure 4. The crushing ratio BM increased as the compaction energy increased.

Figure 5 shows the relationship between the maximum dry density and water absorption under the compaction energy E_4. Water absorption and the maximum dry density have an evident relationship. The maximum dry density decreases as water absorption increases. The maximum dry densities of natural crushed stones for mechanical stabilization, which are generally used as a roadbed material, are around 2 t/m^3. It shows that the density of crushed concrete with the absorption ratio over 3% is still smaller than that of natural materials after sufficient compaction.

Figure 6 shows the relationship between the crushing ratio BM and water absorption under compaction energy E_4. The crushing ratio BM tends to become larger as water absorption increases. The crushing ratio BM of natural crushed stones for mechanical stabilization was about 10% according to the test results in the past. It is thought that the crushed concrete with water absorption of less than 3% has sufficient durability against crushing of particles as natural materials.

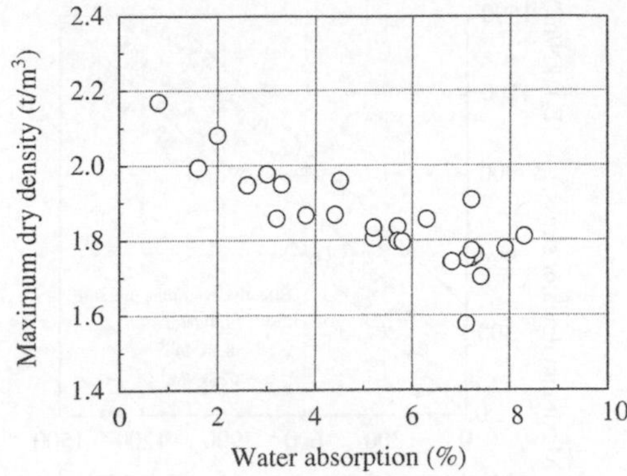

Figure 5. Relationship between maximum dry density and water absorption.

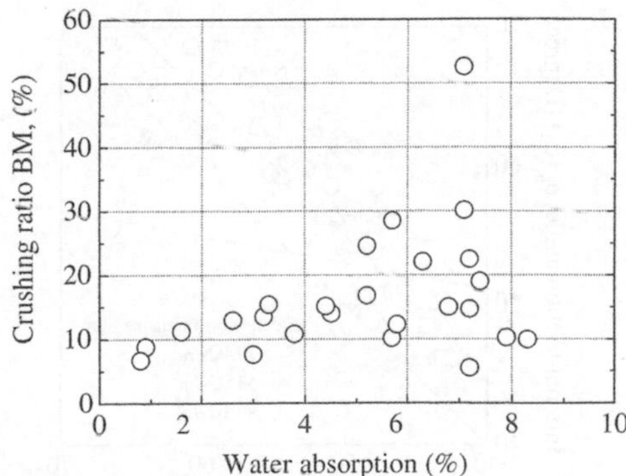

Figure 6. Relationship between crushing ratio BM and water absorption.

4 STRENGTH OF CRUSHED CONCRETE

In order to investigate the strength characteristics of crushed concrete, static triaxial compression tests were carried out under a confined drained condition by using unsaturated specimens. Diameters of specimens were 75 mm, and their heights were 150 mm. Densities of the specimens were prepared to be the same as the maximum dry density in the compaction test under compaction energies E_2 and E_4. Effective confining pressures for triaxial tests were 29.4 kN/m^2, 58.8 kN/m^2 and 88.3 kN/m^2. The axial strain rate was 0.05%/min. Additionally, water-permeated specimens were prepared for the case with compaction energy E4. By using these specimens, triaxial tests were carried out under a consolidated drained condition.

Figure 7 compares the maximum deviator stress q_{max} between the specimens compacted at energies E_2 and E_4. Correspondingly, figure 8 compares the internal friction angle ϕ and figure 9 the residual stress q_r. Here, the residual stress q_r was

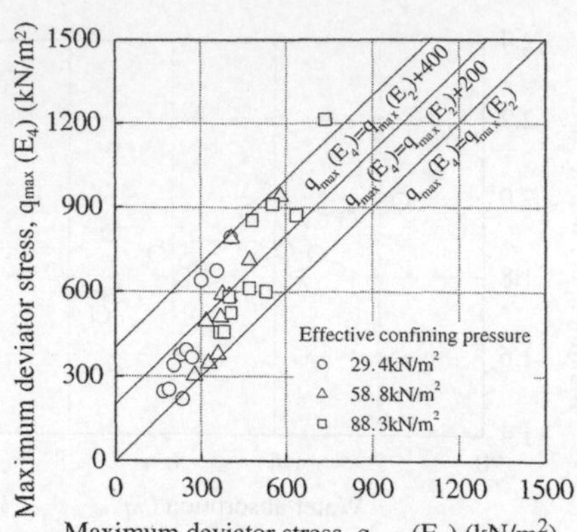

Figure 7. Difference of maximum deviator stress by different compaction energy.

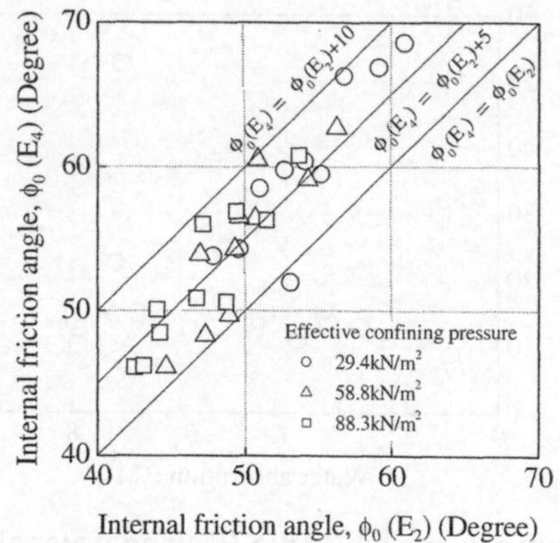

Figure 8. Difference of internal friction angle by different compaction energy.

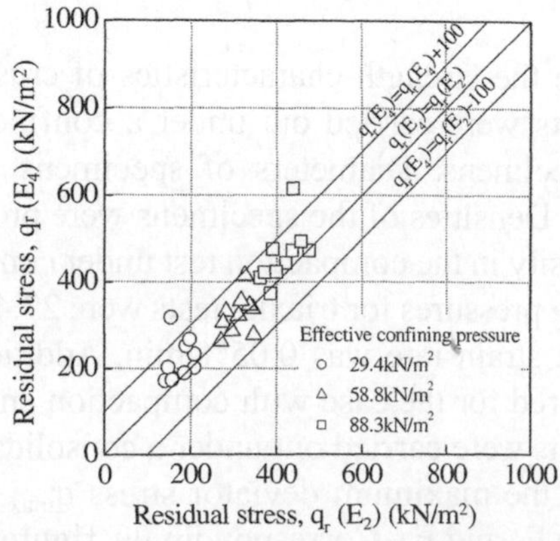

Figure 9. Difference of residual stress by different compaction energy.

defined as the deviator stress at the axial strain of 15%. The internal friction angle was defined by the tangent line of single mohr's circle as:

$$\phi_0 = \sin^{-1}\{(\sigma_1 - \sigma_3)/(\sigma_1 + \sigma_3)\} \tag{2}$$

The specimen compacted by energy E_4 had a larger maximum deviator stress and internal friction angle than the specimen compacted by energy E_2. Concerning the residual stress however, the difference between compaction energies E_2 and E_4 was small. These results show that in larger compaction energies, the maximum deviator stress and internal friction angle become larger, but the compaction energy has little effect on the residual stress.

Then, particle crushing during a triaxial compression test was investigated. Figure 10 compares the crushing ratio BM under compaction energies E_2 and E_4. This figure shows that under smaller compaction energies, crushing ratio BM becomes larger.

Figure 11 compares the maximum deviator stress in the cases with saturated and unsaturated conditions. Correspondingly, figure 12 compares the internal friction angle. The maximum deviator stress in the case of saturated condition tends to become smaller than in the case of unsaturated condition. In a few cases, the decrease in the maximum deviator stress under the saturated condition amounted to 200 kN/m^3. The internal friction angle became also smaller in the case of saturated condition. Furthermore, the residual strength became smaller in the case of saturated condition as shown in figure 13.

Figure 14 compares the crushing ratio BM between saturated and unsaturated conditions. However, the data spread in a relatively wide range, the crushing ratio BM seems to become larger in the saturated condition. It is considered that in the saturated condition, the water absorbed in the particles accelerates deterioration.

Figure 15 shows the relationship between the residual stress ratio and water absorption. The residual stress ratio is defined by dividing the residual stress in the

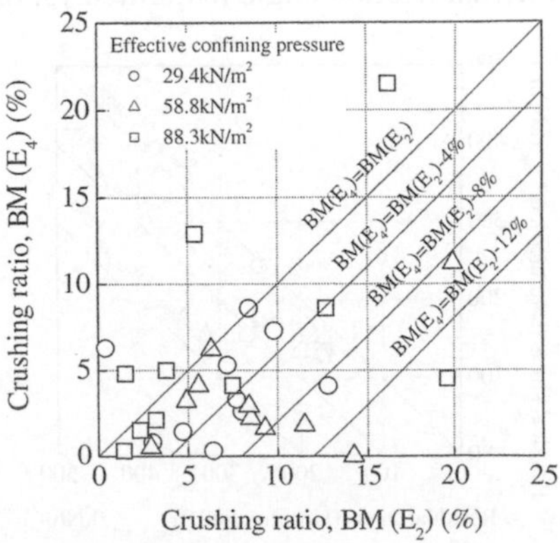

Figure 10. Difference of crushing ratio BM by different compaction energy.

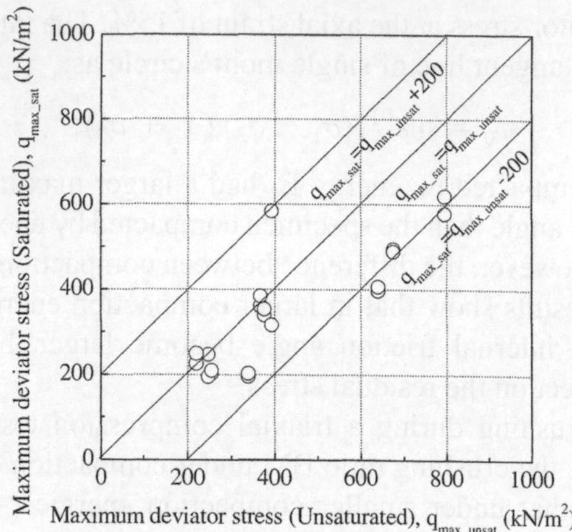

Figure 11. Difference of maximum deviator stress (saturated vs. unsaturated).

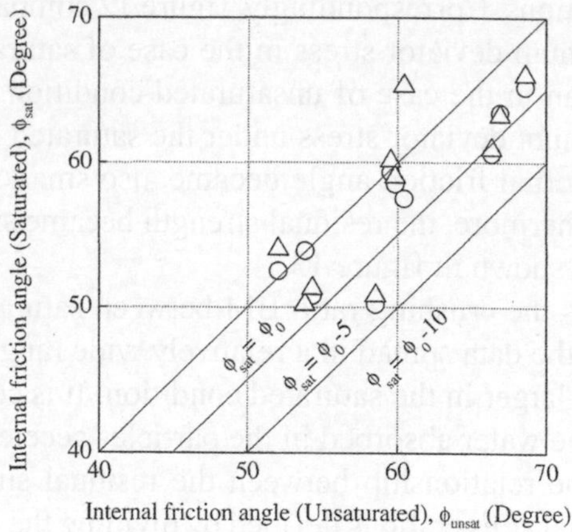

Figure 12. Difference of internal friction angle (saturated vs. unsaturated).

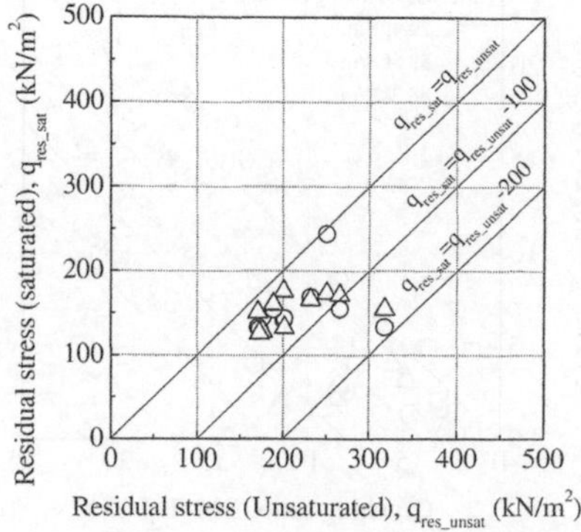

Figure 13. Difference of residual stress (saturated vs. unsaturated).

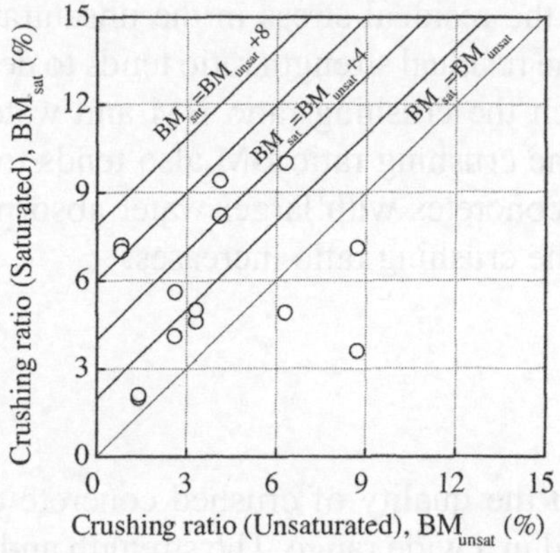

Figure 14. Difference of crushing ratio BM (saturated vs. unsaturated).

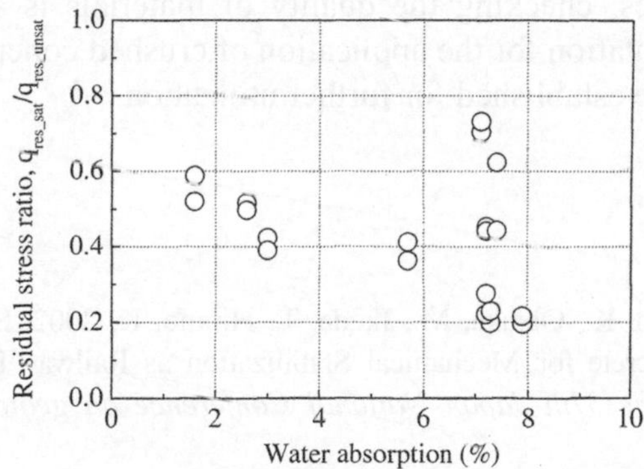

Figure 15. Relationship between residual stress and water absorption.

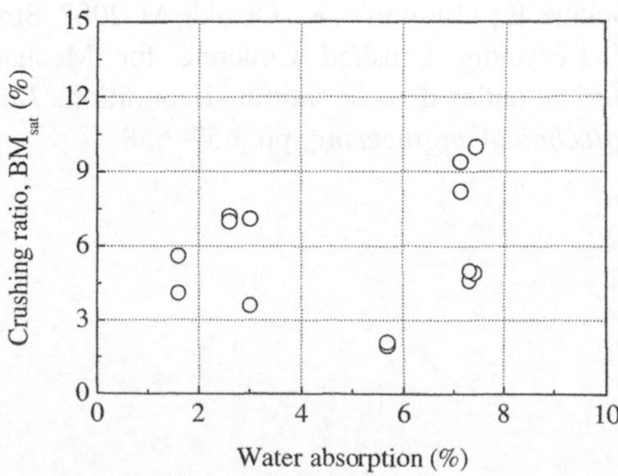

Figure 16. Relationship between crushing ratio BM and water absorption.

saturated condition by the residual stress in the unsaturated condition. As water absorption increases, the residual strength ratio tends to decrease. Figure 16 shows the relationship between the crushing ratio BM and water absorption. As water absorption increases, the crushing ratio BM also tends to increase. These results show that for crushed concretes with larger water absorption, the residual stress tends to decrease and the crushing ratio increases.

5 CONCLUSIONS

This study showed that the quality of crushed concrete collected from different processing plants varied in a wide range. The strength and the compaction characteristics correlated with water absorption of crushed concrete particles. Through the investigation, the effect of water absorption on particle crushing and strength became apparent in a triaxial compression test. To apply crushed concrete to railway earth structures, checking the quality of materials is an important issue. Furthermore, a regulation for the application of crushed concrete to railway earth structures should be established for further utilization.

REFERENCES

Sekine, E., Muramoto, K., Otsuka, M., Ikeda, T., Hirano, K. 2002. Study on Applicability on Crushed Concrete for Mechanical Stabilization as Railway Embankment Material (in Japanese), *The 37th Japan National Conference on geotechnical engineering*, pp. 659–660.

Ikeda, T., Hirano, K., Sekine, E., Muramoto, K., Otsuka, M. 2002. Strength and Deformation Characteristics of Recycling Crushed Concrete for Mechanical Stabilization as Embankment Materials on Unsaturated Condition, (in Japanese), *The 37th Japan National Conference on geotechnical engineering*, pp. 655–656.

Hirano, K., Ikeda, T., Sekine, E., Muramoto, K., Otsuka, M. 2002. Strength and Deformation Characteristics of Recycling Crushed Concrete for Mechanical Stabilization as Embankment Materials under diverse saturated condition, *The 37th Japan National Conference on geotechnical engineering*, pp. 657–658.

Author Index

T - #0231 - 071024 - C0 - 246/174/12 - PB - 9780367389086 - Gloss Lamination